SpringerBriefs in Molecular Science

Electrical and Magnetic Properties of Atoms, Molecules, and Clusters

Series editor

George Maroulis

For further volumes:
http://www.springer.com/series/11647

Sven Heiles · Rolf Schäfer

Dielectric Properties of Isolated Clusters

Beam Deflection Studies

Springer

Sven Heiles
Rolf Schäfer
Eduard-Zintl Institut für Anorganische und
Physikalische Chemie
Technische Universität Darmstadt
Alarich-Weiss-Straße 8, 64287
Darmstadt
Germany

ISSN 2191-5407 ISSN 2191-5415 (electronic)
ISBN 978-94-007-7865-8 ISBN 978-94-007-7866-5 (eBook)
DOI 10.1007/978-94-007-7866-5
Springer Dordrecht Heidelberg New York London

Library of Congress Control Number: 2013951802

Printed on acid-free paper

Springer is part of Springer Science+Business Media (www.springer.com)

Contents

Chapter 1
Introduction

Clusters are agglomerates of several similar atoms or molecules [1, 2]. These objects show size and composition dependent properties, which can not be explained by the well-known properties of the corresponding atoms or solids. Therefore, a variation of the size and composition of these clusters allows the fabrication of small aggregates with interesting and new material properties for possible future applications, like nano-electronics, nano-optics or catalysis [3, 4].

Particularly the dielectric properties of clusters are of major importance, since they determine cluster-cluster interactions, the optical properties and are a probe for the geometric and electronic structure. In order to rule out the influence of the surrounding (surface, ligands), which might affect the dielectric behaviour, the best way is to study clusters isolated in the gas phase. Since the purpose of the experiments is to investigate the size dependent evolution of these properties, a size sensitive method is required. Furthermore, it has to be taken into account, that several structural isomers can exist and their number is dramatically increasing with cluster size and compositional complexity. Therefore, isolated and mass selected clusters have to be analyzed by structure sensitive methods in order to study the dielectric properties. The best way to determine the dielectric behaviour is to measure the influence of an electric field on the potential energy of a neutral particle, i.e. the Stark effect [5–8]. The investigation of the Stark effect of clusters is done with molecular beam electric deflection experiments [9] and offers an universal approach to determine the dielectric properties of neutral particles in the gas phase as a function of size and composition.

The dielectric properties can be divided into two parts. The first is the so-called induced electric dipole moment which is a function of the applied electric field strength. The second part is field independent and is known as the permanent electric dipole moment. Both can be used as sensitive probes for the geometrical, as well as for the electronic structure of a cluster [8]. The induced dipole moment results from the field induced distortion of the electron density and it manifests in the strength of the electronic polarizability. The polarizability α is connected to the electronic energy levels of the corresponding molecules or clusters [8]. Therefore, it is sensitive

S. Heiles and R. Schäfer, *Dielectric Properties of Isolated Clusters*, SpringerBriefs in Electrical and Magnetic Properties of Atoms, Molecules, and Clusters,
DOI: 10.1007/978-94-007-7866-5_1, © The Author(s) 2014

to the electronic structure of the investigated particles. This becomes clear when considering a 2-state model. In this model the polarizability depends on the transition dipole moment μ_{01} connecting the two states with energy ϵ_0 and ϵ_1 [10]

$$\alpha = \frac{2\,|\mu_{01}|^2}{\epsilon_1 - \epsilon_0}. \tag{1.1}$$

In the one-electron approximation these two states are mainly associated with the highest occupied molecular orbital (HOMO) and the lowest unoccupied molecular orbital (LUMO), indicating the sensitivity of the polarizability on the electronic structure of the cluster. For clusters for which the dielectric response is exclusively due to the field induced part of the dipole moment, i.e. non-polar clusters, the value of the polarizability is in a straight forward fashion obtained from the deflection of the clusters in an electric field. The size and composition dependence of the dielectric and electronic properties of non-polar clusters is, therefore, easily derived from electric deflection experiments. The obtained values of the polarizability are valuable benchmarks to test the accuracy of theoretical calculations, i.e. electronic structure calculations employing quantum chemical methods [14, 15].

This behaviour is to a very good approximation found for sodium clusters, for which the magnitude of the permanent dipole moments and consequently the influence on the beam deflection is very small. In Fig. 1.1 the polarizabilities per atom α/N are shown as a function of the cluster size N.[1] The values of α/N have been obtained from the deflection of a molecular beam in an inhomogeneous electric field [11]. Distinct variations are observed which correspond to spherical shell closings, i.e. in dependence of the valence electron count (= number of sodium atoms N in the cluster) degenerate molecular orbitals are completely filled with electrons. This results in minima and maxima of the polarizability indicating the exceptional electronic structure of clusters with shell closings. The overall decrease of the polarizability is explained by representing the sodium clusters as a conducting sphere with a spill-out enhanced radius. This explanation is also supported by density functional calculations using the spherical jellium model (SJM) [12]. The observations for sodium clusters highlight that clusters having 200 atoms still exhibit different dielectric properties when compared to bulk sodium.

For clusters that possess a permanent dipole moment, the analysis of the electric deflection experiments is much more demanding. The permanent dipole moment $\boldsymbol{\mu}_0$ results from an imbalanced distribution of the electrons around the charged nuclei of the particle. It depends for neutral clusters on the distances of the electrons and nuclei with respect to an arbitrarily chosen origin

$$\mu_0 = \int \mathbf{r} p(\mathbf{r}) d\mathbf{r} + \sum_M \mathbf{R}_M Z_M. \tag{1.2}$$

---

[1] In order to convert the polarizability from Å^3 (most commonly used in this book) to SI units (Cm^2/V) or atomic units (a.u.), the values have to be multiplied by $4\pi\epsilon_0 10^{-30}$ or $10^{-30} a_0^{-3}$, respectively. Here, a_0 is the Bohr radius and ϵ_0 is the vacuum permittivity.

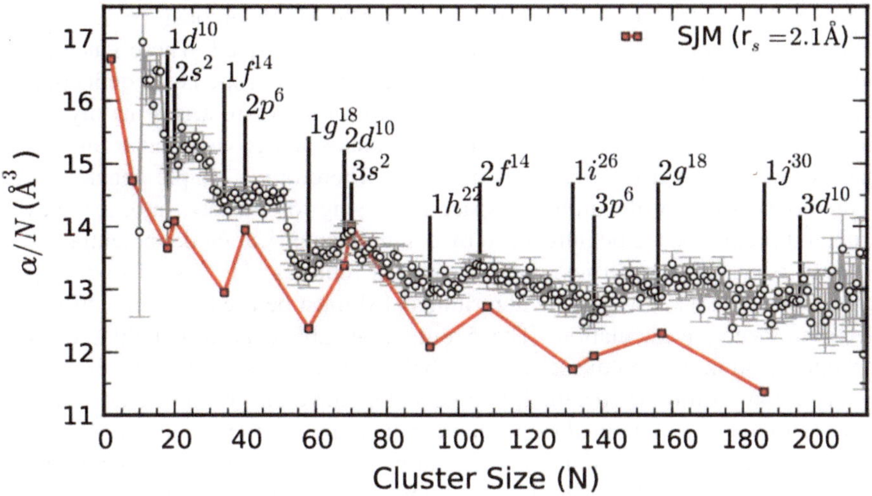

Fig. 1.1 Polarizability per atom α/N for Na_N ($N = 10$–200) [11]. The observed behaviour is explained by treating the electrons as independent particles, which are confined to a spherical box. The radius of the box is given by the number of atoms and the density of bulk sodium. Shell closings have been marked by specifying the principal and the angular momentum quantum number. They qualitatively coincide with extrema observed in experiment, while the overall decay of the experimental polarizability agrees with the conducting sphere model. The cluster Na_{200} surprisingly still exhibits a significantly increased α/N value when compared to a polarizability value of 9.4 Å^3 of a metallic sphere with bulk sodium properties. Shown for comparison are the predicted polarizabilities from spherical jellium model (SJM) calculations [12] taking a Wigner Seitz radius $r_s = 2.1$ Å [13] into account. Reprinted figure with permission from Bowlan et al. [11]. Copyright 2011 by the American Physical Society

In Eq. 1.2, $p(\mathbf{r})$ represents the electron density which depends on the position vector $\mathbf{r}$ and $\mathbf{R}_M$ are the position vectors of the nuclei with charge $\cdot\, Z_M$. If the nuclei of the particle can be treated as rigid, the permanent electric dipole moment is closely connected to the geometrical structure of the cluster. In particular, the dipole moment with respect to the shape of the cluster, described by the moments of inertia, affects the Stark effect. Therefore, the knowledge of the cluster's geometry is necessary in order to analyze the electric beam deflection experiments and extract permanent dipole moments. If, on the other hand, a theoretically predicted geometrical isomer is able to reproduce the experimentally measured Stark effect, beam deflection measurements allow to obtain cluster sensitive information. This enables discussion of size and composition dependent dielectric properties of clusters based on their geometric structure. The ultimate goal of these measurements is, therefore, to answer the question: How do dielectric properties of solids emerge during the growth of the clusters from atoms? It is the major aim of this review to explain the principles of this methodology in detail and to demonstrate that this approach is indeed possible and practical in addition to answering the previously posed question.

In extension to the experimental considerations, theoretical predictions of structural isomers with lowest energy will also be described. These calculations are far from trivial and necessary if an in-depth analysis of the beam deflection results is performed. The difficulty in finding cluster structure arises mainly due to the fact that clusters are able to form chemical bonds and geometrical arrangements that are not known from molecules or solids. Therefore, a simple general model of the connectivity of the atoms in space is missing for clusters. In order to avoid structural predictions to be influenced by common empirical model potentials, such as those from bulk properties, unbiased search routines like Basin-Hopping [16], coalescence kick [17] or genetic algorithms [18] must be used in conjunction with electronic structure calculations. These routines are able to perform unbiased geometry optimizations and predict putative global minimum cluster structures if the atom type and composition is specified. In a next step, the dielectric properties for these cluster structures are obtained from accurate quantum chemical calculations and the Stark effect for each individual isomer can be calculated. In the following, experimental and theoretical aspects for the determination and calculation of the Stark effect are discussed. It will be demonstrated that a careful analysis of experimental data with respect to quantum-chemically predicted structural isomers allows to extract the geometric structure and the dielectric properties of isolated clusters in the gas phase.

This book is organized in such a way that experimental and theoretical aspects are described in detail, so that the experimental results discussed at the end of the book can easily be understood. Therefore, an experimental realization of the molecular beam electric deflection method is first introduced. This is followed by an in-depth theoretical description and modeling of the Stark effect and a discussion of the capabilities of using the Stark effect to discriminate between different cluster isomers. This includes the description of unbiased structure search routines [19], a brief explanation of quantum chemical methods and the modeling of the Stark effect by classical and quantum mechanics. Then, in the next section various applications of the technique to molecular clusters and complexes, to metal and semiconductor clusters of group 14 and to core-shell clusters and nanoalloys are presented. It is shown how molecular beam electric deflection experiments in combination with quantum chemistry are used to extract geometrical and dielectric properties of these complexes and clusters. In the final chapter, two novel experimental tools are described, which are also based on the Stark effect, allowing to determine frequency-dependent dielectric properties and to manipulate the motion of large neutral molecule and clusters in the gas phase. A short summary finishes the considerations.

References

1. Haberland H (1995) Clusters of atoms and molecules I. Springer-Verlag, Berlin
2. Bergmann L, Schaefer C (1992) Experimentalphysik 5: Vielteilchen-Systeme. Walter de Gruyter Verlag, New York
3. Sattler KD (2010) Handbook of nanophysics: nanoelectronics and nanophotonics. Taylor & Francis, Boka Raton
4. Freund HJ, Meijer G, Scheffler M, Schlögl R, Wolf M (2011) Angew Chem Int Ed 50:10064
5. Stark J (1913) Nature 92:401
6. Stark J (1914) Ann Phys 348:965
7. Scheffers H, Stark J (1934) Phys Z 35:625
8. Bonin K, Kresin VV (1997) Electric-dipole polarizabilities of atoms, molecules and clusters. World Scientific Publishing Company, Singapore
9. Ramsey NF (1956) Moelcular beams. Oxford University Press, Oxford
10. Landau LD, Lifschitz EM (2007) Lehrbuch der Theoretischen Physik: Quantenmechanik. Verlag Harri Deutsch, Frankfurt am Main
11. Bowlan J, Liang A, de Heer WA (2011) Phys Rev Lett 106:043401
12. Ekardt W, Penzar Z (1986) Solid State Commun 57:661
13. Ashcroft NW, Mermin DN (2007) Festkörperphysik. Oldenburg Verlag, München
14. Schäfer S, Mehring M, Schäfer R, Schwerdtfeger P (2007) Phys Rev A 76:052515
15. Thierfelder C, Assadollahzadeh B, Schwerdtfeger P, Schäfer S, Schäfer R (2008) Phys Rev A 78:052506
16. Wales DJ, Doye JPK (1997) J Phys Chem A 101:5111
17. Saunders M (2004) J Comput Chem 25:621
18. Johnston RL (2003) Dalton Trans 4193–4207.
19. Heiles S, Johnston RL (2013) Int J Quantum Chem 113:2091

Chapter 2
Molecular Beam Electric Field Deflection: Experimental Considerations

To understand the electric beam deflection method and the possibility to extract the dielectric properties of isolated clusters from these measurements, the essential components of the molecular beam apparatus are explained first. The experimental setup of a molecular beam apparatus build for beam deflection studies allows many variations and differs according to the kind of the studied cluster system. However, for the sake of simplicity we will introduce the basic experimental procedure with the help of an apparatus developed in our group. Possible modifications or instrumental variations will be indicated in the corresponding sub-chapters. First, the experimental procedure will be described to give the reader an idea how to experimentally realize these measurements and what components are required. Furthermore, we will phenomenologically introduce the measurement principle by discussing the observations made when performing a beam deflection experiment (Sect. 2.1). It will become clear, that the intensity distribution in the molecular beam differs between the field free measurements and those experimental runs when an electric field is applied. We will see that this difference can only be explained by a deflection of the clusters due to the electric field and that this deflection can qualitatively understood when taking the Stark effect of the cluster into account. A more detailed and quantitative discussion of this observation will be given in Chap. 3. In the subsequent sections the different components of the experiment are discussed in more detail. This includes the necessary vacuum setup (Sect. 2.2), the cluster generation in a cluster source (Sect. 2.3), the electric deflection unit (Sect. 2.4) and the mass spectrometer used to ionize and detect the neutral clusters (Sect. 2.5).

2.1 Experimental Setup and Measurement Principle

A schematic overview of the molecular beam apparatus which is used to study the dielectric properties of isolated clusters is shown in Fig. 2.1. First, the general experimental setup and the measuring principle is described. Important experimental details

S. Heiles and R. Schäfer, *Dielectric Properties of Isolated Clusters*, SpringerBriefs in Electrical and Magnetic Properties of Atoms, Molecules, and Clusters,
DOI: 10.1007/978-94-007-7866-5_2,

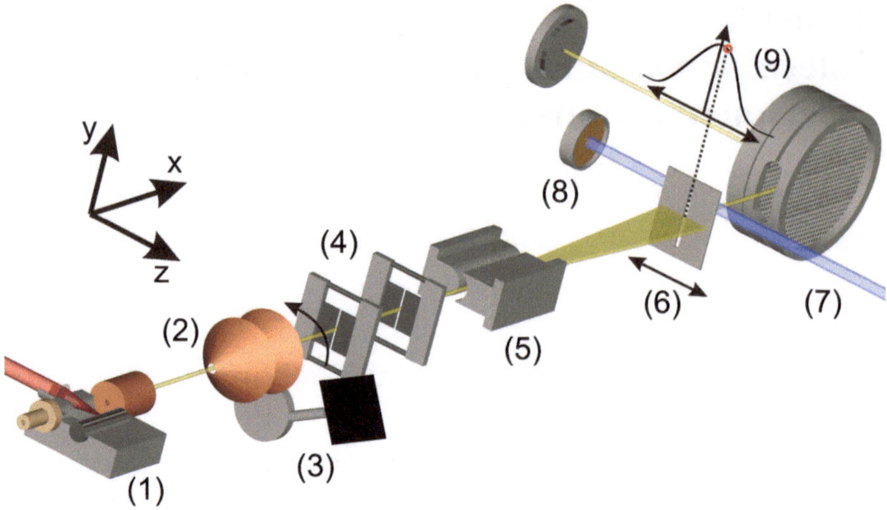

Fig. 2.1 Schematic setup of the molecular beam apparatus developed in our group: Cluster source with cooled expansion nozzle (*1*), double skimmer (*2*), shutter unit (*3*), collimatores (*4*), electric deflection unit (*5*), movable slit aperture (*6*), F_2-Excimer laser (*7*), pyro-electric detector (*8*), time-of-flight mass spectrometer (*9*). The molecular beam is highlighted in *yellow* and the laboratory coordinate system is shown in the *upper left* part of the figure

are discussed in more detail in the following sections. Parts of the experimental setup have been described in references [1–3]. The experiment, which is repeated with a rate of 10 Hz, starts with the production of clusters in a laser vaporization source (1) (see also Sect. 2.3). Clusters are formed in a helium atmosphere after a laser pulse hits the target and are subsequently expanded through a nozzle into high vacuum. The cluster-helium mixture is thereby supersonically expanded and a molecular beam (Fig. 2.1, yellow) is formed [4], which is narrowed by two skimmers (2) with circular openings of 2 and 3 mm, respectively. In a molecular beam all particles exhibit a directed translation motion.[1] As a consequence of this directed motion and the absence of collision processes due the low pressure (see also Sect. 2.2), the clusters can be considered as isolated particles [4]. The molecular beam passes a home-built shutter unit (3) [5]. At this point, it is possible to interrupt the particle beam at a well defined point in time, in order to measure the flight time of the clusters. The additional knowledge of the flight path (3.38 m) enables the determination of the particle speed v with an accuracy of (2–3) %. Next, the molecular beam passes a double collimator unit (4) resulting in a collimated 0.50-mm-wide (z-direction) and 3.00-mm-long (y-direction) rectangular beam shape. This rectangular beam enters the electric deflection unit (5) which will be described in Sect. 2.4. After leaving of the deflection unit (5) the clusters enter a 1.59-m-long area of free flight.

[1] This is in contrast to a thermal velocity distribution for which $< v_x > = < v_y > = < v_z >$ [4].

Subsequently, a, in z-direction, movable slit (6), which is 0.35 mm in width and 20.00 mm in height, is reached. The aperture exclusively allows a small portion of the molecular beam to proceed towards the ionization laser. These clusters are ionized by a F_2-Excimer laser (7), whose fluence is monitored during a measurement by a pyro-electric detector mounted in the apparatus (8). The generated cluster ions are detected in a self-built time-of-flight mass spectrometer (9) and a photo-ionization mass spectrum is obtained (Sect. 2.5). Consequently, the described experiment allows the mass-resolved investigation of neutral clusters.

To be able to detect the influence of an inhomogeneous electric field on the clusters experimentally, it is necessary to record cluster intensities as a function of the movable slit position (6) and the electric field strength (5). First the position of the slit aperture is changed keeping the electric field constant. With the help of the before described collimators the molecular beam owns a predetermined intensity distribution along the z-axis, defined in the laboratory coordinate system.

The particle density is not constant along the z-axis, but will exhibit a maximum on the molecular beam axis[2] and will decrease when moving away from this position.[3] This distribution of the particle density as a function of the z-position is called a molecular beam profile $\psi(z)$. Because the slit of the aperture is smaller than the dimensions of the molecular beam, only a small part of the beam is detected in the mass spectrometer. The actual number of clusters which can pass the slit depend on the aperture position. The particle density distribution of the molecular beam and, therefore, ψ can be measured sequentially by shifting the slit along the z-axis. Schematically, a beam profile without electric field (ψ_0, dashed line) is shown in Fig. 2.2c. When applying a voltage to the deflection unit, the molecular beam profile is influenced by the electric field. This is schematically shown in Fig. 2.2c. In this example, the beam profile with field ψ_1 (black) is shifted and broadened

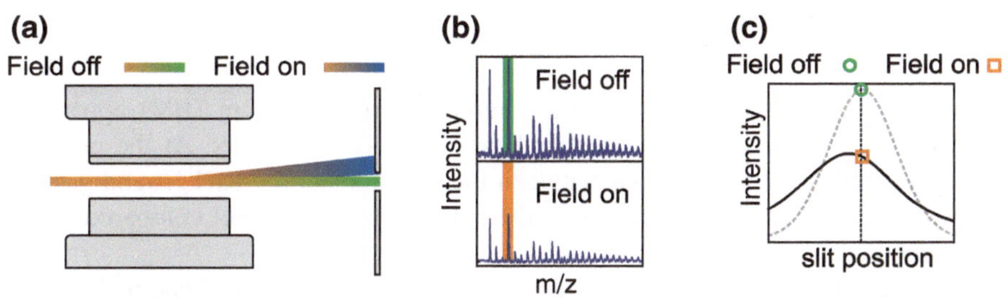

Fig. 2.2 **a** Schematic trajectory of the molecular beam with switched off and switched on electric field. **b** Mass spectra obtained for the two experiments shown in **a**. **c** Integrated signal intensity with the electric field switched off and on, respectively. The corresponding beam profiles ψ_0 (*gray* sketched) and ψ_1 are shown (*black*), too

[2] This position is chosen to be z_0. Hence, it is convenient to use the relative z-scale $p = z - z_0$.

[3] Furthermore, we note that for this experimental setup the particle density must be symmetric around z_0. The observed intensity distribution can be described with a Gaussian function, what will be done in all what follows.

when compared to ψ_0. In order to qualitatively understand this result, we want to discuss the experimental data acquisition and processing for a single cluster at a given slit position. If no electric field interacts with the clusters in the molecular beam, a photo-ionization mass spectrum is obtained as it is schematically shown in Fig. 2.2b for Si_N clusters (N is the number of atoms in the aggregate). In a next step, the mass spectrometric signal for an arbitrarily chosen cluster, e.g. Si_{11} (highlighted in green), is integrated and is plotted as a function of the slit position. As a consequence, the value of the molecular beam profile ψ_0 at this specific slit position is obtained (green circle in Fig. 2.2c). In the next step, the electric field is switched on and the intensity of the same cluster at the same slit position is measured again. A drop in intensity can be observed in Fig. 2.2b and a modified value for ψ_1 (orange square in Fig. 2.2c) is obtained.[4] The only possible explanation that can rationalize this observation, is a deflection of the clusters in the molecular beam due to the applied electric field (see Fig. 2.2a). When deflected less clusters can pass the movable slit and the signal intensity will decrease. This effect can qualitatively be understood, if the dielectric response of the cluster in the electric field is taken into account. The applied electric field interacts with the electric moments of the neutral cluster (also called dielectric properties) and will influence the energy levels of the cluster, what is known to be the Stark effect [6]. Since the electric field is inhomogeneous (see Sect. 2.4) this interaction will lead to a net force acting on the cluster, resulting in the observed deflection. Therefore, the deflection is a measure of the Stark effect experienced by the cluster and is connected to the cluster's dielectric properties. By rationalizing the beam deflection we can, therefore, extract the dielectric properties from beam deflection experiments. A detailed theoretical description of the Stark effect and the beam deflection is given in Chap. 3.

2.2 Vacuum System

The experiments described in Sect. 2.1 require a high vacuum (HV) apparatus, ensuring the formation of a molecular beam of isolated clusters. At the pressure of below $\sim 10^{-6}$ mbar collisions between the clusters and the background gas can be avoided if the overall length traveled by the particles is small compared to the mean free path[5] [4].

The experimental realization of the vacuum system is shown schematically in Fig. 2.3. In principle four chambers can be distinguished. The pressure realized during operation is described in the following. The laser vaporization source, in which helium pulses are introduced, is located in the source chamber (1). The working

[4] To obtain the whole beam profiles the described experiment with and without field, needs to be repeated for several slit positions.

[5] A rough estimate based on hard sphere collision cross section of N_2 shows that at 300 K and $\sim 10^{-5}$ mbar the mean free path is still about 10 m. Hence, for the described operating conditions collisions can be ruled out.

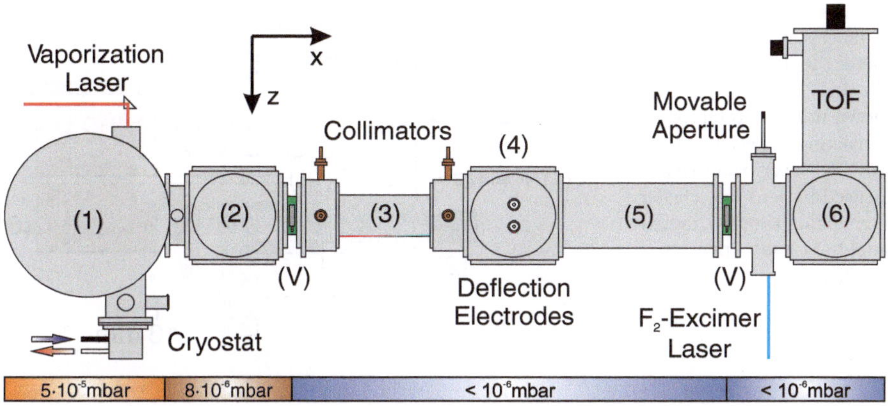

Fig. 2.3 Schematic view from the $+y$ direction on the molecular beam apparatus: Source vacuum chamber (*1*) with Nd:YAG laser beam (*red*, with prism) and He cryostat (*blue* and *red arrow*, cold and heated He stream), first differentially pumped chamber (*2*), flight tube (*3*) with collimators (*bronze-colored*), vacuum chamber for deflection electrodes (*4*), flight tube (*5*), vacuum chamber (*6*) with F_2-Excimer laser beam (*cyan*), slit aperture (*gray-black*) and time-of-flight (TOF) mass spectrometer. The high-vacuum (HV) valves (V) are shown in *green*. The pressure scale below the apparatus indicates the pressure during the experiment

pressure of $\sim 5 \cdot 10^{-5}$ mbar is reached by a diffusion pump (Leybold Heraeus, 12000 L/s) which itself is pre-pumped by a roots pump and an oil rotary vane pump. The source chamber (1) is connected by a double skimmer to the next differentially pumped chamber (2). The vacuum chamber (2) which among other things houses the shutter unit (see Sect. 2.1 and Fig. 2.1) is evacuated by a pumping stage consisting of a rotary vane and a diffusion pump (Leybold Heraeus, 1200 L/s) resulting in a pressure of $\sim 8 \cdot 10^{-6}$ mbar under operating conditions. The next vacuum chamber, in which the double collimator (3) and the deflection unit (4) as well as the first flight tube is located (5), can be separated from chamber (2) and (6) by means of two HV valves (V, *green*). The whole vacuum chambers (3), (4) and (5) are equipped with a rotary vane pump in combination with a diffusion pump (Edwards Diffstack 250/2000 m, 2000 L/s) achieving a base pressure below 10^{-6} mbar. The last vacuum chamber (6) contains the movable slit aperture, a MgF_2 window which allows the F_2-Excimer laser light to enter the vacuum chamber and the time-of-flight mass spectrometer. A pressure of less than 10^{-6} mbar is generated by a turbo molecular pump (Varian turbo-V 3K-T, 2050 L/s), pre-evacuated by a rotary vane pump.

2.3 Cluster Source Design

For the generation of clusters a whole series of sources have been developed. Beside sputter [7] and gas aggregation sources [8, 9], the laser vaporization technique is of special importance because clusters of very hard, non-conducting but also conducting

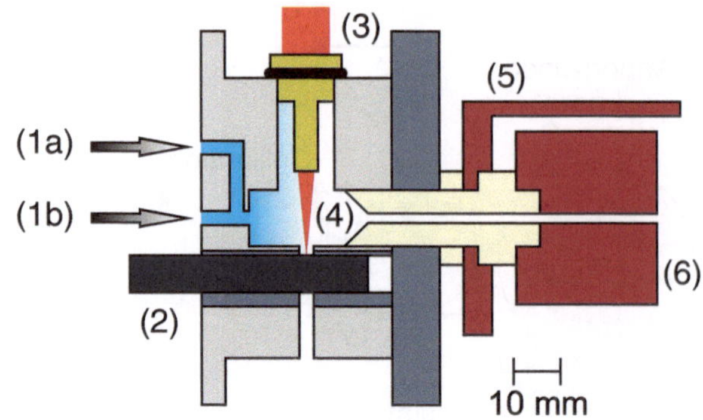

Fig. 2.4 Cross section of the laser vaporization source: Inert gas entries (*1a* and *1b*, inert gas shown in *blue*), rotating and translating material rod (*2*), Nd:YAG-laser beam (*red*) focused by lens (*3*), cluster aggregation zone (*4*), radiation shield (*5*) and expansion nozzle (*6*)

materials can be generated. Therefore, most of the deflection experiments have been performed using laser vaporization sources and we will describe the source developed and used in our group in more detail.

The cluster source, which is shown in Fig. 2.4, is a home-built laser vaporization source based on the design of Smalley [10] and deHeer [11]. For the generation of clusters, helium, with a background pressure of ~8 bar, is injected through of a pulsed valve (typical opening times 350–1000 μs) in the source via the openings (1a) or (1b). The variable helium flow permits to modify the dwell time of the helium in the source. If gas enters through (1a) the formation of small clusters is preferred, while the use of (1b) is benificial for the formation of larger aggregates. Typically 400–800 μs after the helium pulse a Nd:YAG (yttrium aluminum garnet, pulse length 10 ns @ 1064 nm) laser with an energy between 40 and 100 mJ per pulse is focused by a lens (3) on a rotating and translating material rod (2). A few mono-layers of the material (2) are ablated by the intense laser pulse and a plasma is formed [12], which is rapidly cooled in the helium atmosphere, followed by the formation of clusters (4) by three-body collisions [7]. The pressure of some mbar helium in the source and the background pressure of $5 \cdot 10^{-5}$ mbar in the source chamber results in a supersonic expansion of the cluster-helium mixture through the nozzle with a 2 mm orifice and a total length of 61 mm (6). The first 36 mm of the nozzle is made of Teflon™, while copper was used for the remaining 25 mm. In the copper block a heating element (max. 20 W) is integrated which is controlled by a PID regulator (proportional-integral-derivative, LakeShore 325). Plates made from oxygen-free high thermal conductivity cooper have been used to connect the copper block of the nozzle with a Helium closed cycle cryostat (Sumitomo Heavy Industries, 1 W @ 4.2 K). The temperature of the Teflon™ part of the nozzle remains close to room temperature,[6] while the temperature of the copper block can be varied between 12 and 350 K. To minimize heat looses by radiation, a copper heat shield (5) surrounding the nozzle is precooled by the first cooling stage of the helium cryostat, which is held at 75–300 K. As a consequence of the high pressure in the source, the cluster-helium mixture frequently collides with the nozzle wall [7] and, therefore,

[6] Hence, the cluster source and the cooled part of the nozzle are thermally isolated.

thermalizes with the copper nozzle. After supersonic expansion, the temperatures of the different degrees of freedom of the clusters in the molecular beam are not longer in thermal equilibrium [4, 7]. Thereby, the rotational temperature (T_{rot}) can be lower than the nozzle temperatures, since the rotational degrees of freedom are cooled more effectively by cluster-wall and cluster-gas collisions (see [4, 7, 13, 14] for an in-depth discussion of supersonic expansions and the observed rotational, vibrational and translational temperatures and [15] for an example of typically observed rotational temperatures in supersonic jets). However, the vibrational temperature (T_{vib}) of the clusters in the molecular beam stays close to the nozzle temperature [16]. Consequently, by changing the nozzle temperature, the thermal excitation of the cluster skeleton can be influenced. For very low temperatures clusters can be considered as rigid, while at high nozzle temperatures clusters are thermally excited. These two limiting cases and the influence on the interpretation of the experimental results will be discussed in Chap. 3.

2.4 Deflection Unit

The determination of the electric susceptibilities of clusters requires an inhomogeneous electric field. In general static or dynamic fields can be used. First we will focus only on static deflection units. A more detailed description of the determination of dynamic dielectric properties with the help of time-dependent fields will be given in Sect. 5.1. For the generation of a static electric field, in principle, various electrode geometries can be used. Nevertheless, the by far most often used electrode geometry generates a so called "two-wire" field and we will focus our discussion on this electrode geometry [17, 18]. However, a couple of other electrode geometries have been used to perform deflection studies and some examples can be found in [19, 20].

The electrode geometry that generates an electric "two-wire" field and the corresponding coordinate system are shown in Fig. 2.5a. In analogy to the magnetic "two-wire" field [21] a convex electrode with radius a and a concave electrode with radius b are used. The z-component of the electric field strength is given by

$$E_z = \frac{K}{\sqrt{(a-y)^2+z^2}\sqrt{(a+y)^2+z^2}}, \tag{2.1}$$

and the corresponding field gradient

$$\frac{\partial E_z}{\partial z} = \frac{2 \cdot K \cdot z \cdot (a^2+y^2+z^2)}{\left[(a-y)^2+z^2\right]^{3/2}\left[(a+y)^2+z^2\right]^{3/2}} \tag{2.2}$$

can be obtained by deriving with respect to z. The parameter K present in Eqs. 2.1 and 2.2 is determined by the electrode radii $a = 3.8$ and $b = 4.0$ mm, as well as the potential difference U between the two electrodes [1]

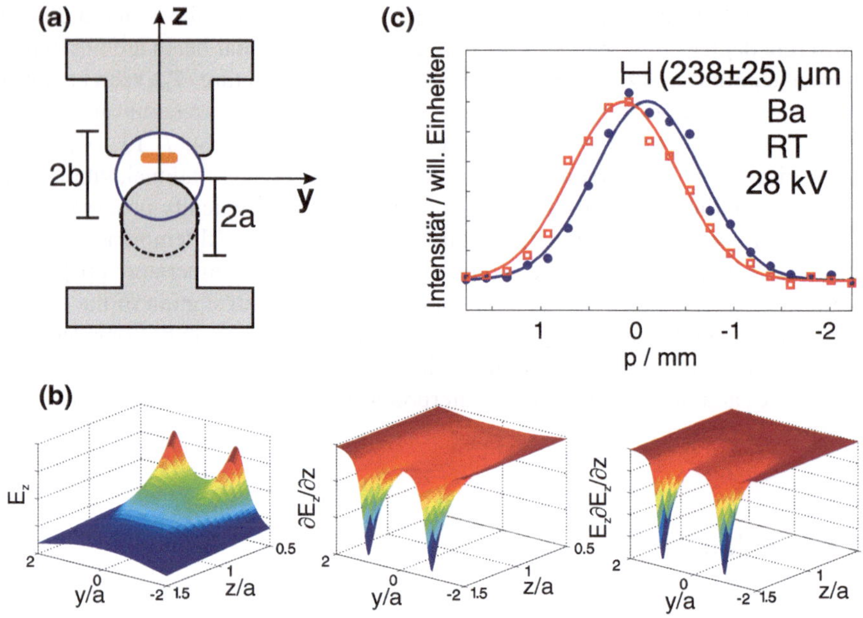

Fig. 2.5 **a** View from $+x$ direction on the deflection electrodes with a "two-wire" geometry. The molecular beam axis is highligthed in *orange*. The diameter of the concave and convex electrodes are given by *2a* and *2b*, respectively. **b** Electric field strength (E_z), field gradient ($\partial E_z/\partial z$) and product of both quantities ($E_z \cdot \partial E_z/\partial z$) for the "two-wire" field geometry in arbitrary units as a function of y/a and z/a. **c** Typical experimental result for the beam deflection of barium atoms (speed ~1420 m/s, *blue dots* without field, *red squares* with field, Gaussian functions with the corresponding color are fitted to the data points) at room temperature (RT) and an applied deflection voltage of 28 kV. The position of aperture is abbreviated by p. A deflection of (238 ± 25) μm was observed

$$U = \frac{K}{a}\left[\arctan\left(\frac{a}{b}\right) - \frac{\pi}{4}\right]. \tag{2.3}$$

A special feature of the "two-wire" field is that the gradient $\partial E_z/\partial z$ as well as the product of the electric field E_z with the electric field gradient, which are responsible for the effects described in Sect. 2.1, remain nearly constant over the spatial dimensions of molecular beam (orange, Fig. 2.5a). In Fig. 2.5b E_z, $\partial E_z/\partial z$ and $E_z \cdot \partial E_z/\partial z$ are shown as a function of y/a and z/a. The position of the molecular beam relative to the electrodes is indicated in Fig. 2.5a in orange. Typically the molecular beam is located at $(1.1 - 1.2)a$ in z- and $(-0.35 - 0.35)a$ in y-direction for which $\partial E_z/\partial z$ as well as $E_z \cdot \partial E_z/\partial z$ change only about ~5 %. Therefore, the force experienced by the particles is to a good approximation independent of their position in the molecular beam (see Eq. 3.3). In order to quantify the observed beam deflection it is necessary to introduce an apparatus constant (Eq. 2.4). It turns out to be useful to introduce the constant

$$\gamma = \frac{\sigma}{U^2} E_z = \frac{l_1^2 + l_1 l_2}{U^2} \frac{\partial E_z}{\partial z} E_z, \tag{2.4}$$

which is independent of the applied deflection voltage. The quantities l_1 and l_2 in Eq. 2.4 are the electrode length (150 mm) and the distance of free flight (1590 mm, see Sect. 2.1), respectively. Furthermore, the constant σ is introduced which depends on U but often it is more convenient to use σ when discussing the beam deflection or the force experienced by particles (see Eqs. 3.3 and 3.5). The calibration constant can be determined by measuring the beam deflection of a particle with well characterized dielectric properties. For our experimental setup we used the barium atom [22, 23]. A typical beam profile obtained for Ba is shown in Fig. 2.5c. The average of five measurements gave a value of $\gamma = (2.8 \pm 0.2) \cdot 10^7\,\text{m}^{-1}$ in very good agreement with theoretical predictions taking Eqs. 2.1 and 2.2 into account [1].

2.5 Position-Sensitive Mass Spectrometry

For the size selected detection of clusters a mass spectrometer (MS) is required. In principle all types of mass spectrometers can be used to perform beam deflection measurements. Nevertheless, in practice mostly time-of-flight (TOF) mass spectrometers are used but also several studies with quadrupole MS [8, 19] are known. The advantage of a TOF-MS is that all cluster species in a molecular beam can be probed at the same time. If the molecular beam profiles are recorded serially by scanning the molecular beam with the help of a aperture, a TOF-MS with space focusing is used [24]. In our setup the neutral clusters pass the aperture and are subsequently ionized by a F_2-excimer laser (wavelength 157 nm, see Fig. 2.1). The laser intensity profile shows a rectangular cross section from $20 \times 10\,\text{mm}^2$ with an average energy per pulse of $150\,\mu$J. The pulse energy can be measured during the experiment by a pyro-electric detector (see Fig. 2.1). The ionized clusters enter the acceleration region of the TOF-MS, in which the clusters are deflected orthogonal to their original flight direction by applying voltages of 3–4 kV in about hundred nanoseconds by high voltage switches. After a 35 cm-long region of free flight the clusters are post-accelerated on an Even cup [25]. The secondary electrons generated are converted in photons by a fluorescence plate. The photons are detected and multiplied by photomultiplier stage. Typically a mass resolution of better than 100 is reached. If a TOF-MS with a reflectron is used the mass resolution could be further enhanced [26].

For the determination of the molecular beam profiles ψ_0 and ψ_1, about 60 experiments are performed at 20 randomly selected slit positions. This means at each slit position on average three experiments are performed. One experiment consists of measuring the photo-ionization mass spectrum with and without electric deflection field at a given slit position, while each mass spectrum is the average of 100–200 pulses.

As an alternative to serially scanning the molecular beam, position sensitive TOF-MS techniques have been used [27–29]. In this approach, the molecular beam is not scanned with the help of aperture, but the molecular beam profile is reconstructed from the flight time. For that purpose the space focusing conditions must be slightly altered, so that different starting positions manifest in different flight times, i.e. a molecular beam deflection and broadening manifests in a deflection and broadening of the time-of-flight signal, respectively. The big advantage of this method is that molecular beam profiles are measured at the same time for all positions without scanning a slit aperture, thereby increasing the signal-to-noise ratio and decreasing the duration of the experiment. However, the serial scanning mode is, in comparison to the position sensitive TOF-MS, more sensitive to small deflections, allowing the investigation of clusters with a small Stark effect or with very high mass. Whether position sensitive detectors will unify the advantages of both operation modes, remains open to future investigations [30].

References

1. Schäfer S (2008) Der Stark-Effekt als Werkzeug zur Strukturaufklärung isolierter cluster. Dissertation, Fachbereich Chemie, TU Darmstadt.
2. Mehring M (2007) Elektrische Molekularstrahlexperimente am Beispiel kleiner neutraler Bariumspezies. Diplomarbeit, Fachbereich Chemie, TU Darmstadt
3. Rohrmann U (2008) Magnetisches Verhalten reiner und Mangan-dotierter Zinn-Cluster. Master's thesis, Fachbereich Chemie, TU Darmstadt
4. Scoles G (1988) Atomic and molecular beam methods, vol 1. Oxford University Press, New York
5. Scholten RE (2007) Rev Sci Instrum 78:026101
6. Scheffers H, Stark J (1934) Phys Z 35:625
7. Haberland H (1995) Clusters of atoms and molecules I. Springer, Berlin
8. Moro R, Rabinovitch R, Xia C, Kresin VV (2006) Phys Rev Lett 97:123401
9. Carrera A, Mobbili M, Marceca E (2009) J Phys Chem A 113:2711
10. Dietz TG, Duncan MA, Powers DE, Smalley RE (1981) J Chem Phys 74:6511
11. Milani P, de Heer WA (1990) Rev Sci Instrum 61:1835
12. Bergmann L, Schaefer C (1992) Experimentalphysik 5: Vielteilchen-Systeme. Walter de Gruyter Verlag, NewYork
13. Mate B, Tejeda G, Montero S (1998) J Chem Phys 108:2676
14. Montero S (2013) Phys Fluids 25:056102
15. Lenzer T, Bürsing R, Dittmer A, Panja SS, Wild DA, Oum K (2010) J Phys Chem A 114:6377
16. Pokrant S (2000) Phys Rev A 62:051201
17. Ramsey NF (1956) Moelcular beams. Oxford University Press, Oxford
18. Pauly H (2000) Atom, molecular and cluster beams II. Springer, Berlin
19. Sievert R, Cadez I, Van Doren J, Castleman AW (1984) J Phys Chem 88:4502
20. Imura K, Ohoyama H, Kasai T (2004) Chem Phys 301:183
21. Rabi II, Millman S, Kusch P, Zacharias JR (1939) Phys Rev 55:526
22. Schäfer S, Mehring M, Schäfer R, Schwerdtfeger P (2007) Phys Rev A 76:052515
23. Schwartz HL, Miller TM, Bederson B (1974) Phys Rev A 10:1924
24. Wiley WC, McLaren IH (1955) Rev Sci Instrum 26:1150
25. Bahat D, Cheshnovsky O, Even U, Lavie N, Magen Y (1987) J Phys Chem 91:2460
26. Mamyrin B (2001) Int J Mass Spectrom 206:251
27. de Heer WA, Milani P (1991) Rev Sci Instrum 62:670
28. Knickelbein MB (2001) J Chem Phys 115:5957
29. Antoine R, Rayane D, Allouche AR, Aubert-Frecon M, Benichou E, Dalby FW, Dugourd P, Broyer M, Guet C (1999) J Chem Phys 110:5568
30. Rahim MAE, Antoine R, Arnaud L, Barbaire M, Broyer M, Clavier C, Compagnon I, Dugourd P, Maurelli J, Rayane D (2004) Rev Sci Instrum 75:5221

Chapter 3
Molecular Beam Electric Field Deflection: Theoretical Description

After having explained the basic experimental components and the measurement principle in Chap. 2 a qualitative and quantitative description of the experimental observations for beam deflection studies employing the two-wire field setup is developed in this chapter. First, the physics behind the observed deflection, which is connected to the dielectric properties and consequently to the Stark effect, of the clusters in a molecular beam is discussed. Second, from these basic considerations it will become apparent that beside some experimental parameters only one molecular quantity needs to be known in order to quantitatively describe the deflection behavior of a rotating object in an electric field. This quantity will be called dipole moment distribution function and depends on various parameters of the deflected particle, e.g. the permanent electric dipole moment and the rotational temperature. Hence, in most of the sections of this chapter different models and approaches are introduced and discussed which aim to calculate the dipole moment distribution function. First, an atomistic model of the cluster structure needs to be developed and its dielectric properties are inferred from quantum chemical calculations. The basic concept of these computations will be described in Sect. 3.2. In Sect. 3.3 the characteristics of the distribution function are derived from perturbation theory (PT) considerations, i.e. in the limit of a particle interacting only weakly with the applied electric field. Thereafter, a description of the classical (Sect. 3.4) and quantum mechanical (Sect. 3.5) treatment of a rotor with variable shape in an electric field of arbitrary field strength follows. All models developed up to this point make use of the rigid rotor assumption and hence the moment of inertia of the particle does not change with time. In Sect. 3.6 the case of a "floppy" cluster which undergoes fast vibrational motions and/or isomerizations on the time scale of the deflection experiment is discussed.

3.1 Particles in an Inhomogeneous Electric Field: Force and Deflection

A good starting point in order to understand the deflection of a beam of neutral clusters by an inhomogeneous electric field is to solve the problem for a single particle, indexed i, first. The first step is to separate the translational motion from the

S. Heiles and R. Schäfer, *Dielectric Properties of Isolated Clusters*, SpringerBriefs in Electrical and Magnetic Properties of Atoms, Molecules, and Clusters,
DOI: 10.1007/978-94-007-7866-5_3, © The Author(s) 2014

internal vibrations and rotations. While the rest of the chapter deals with the adequate description of the rotational and partially the vibrational motion, we start with only considering the translation. We will assume that the translation of the cluster in a molecular beam can be treated classically. Additionally, the following considerations are only valid for the "two-wire" field geometry introduced in Sect. 2.4 and the choice of the coordinate system refers to Fig. 2.1. In this special case, the experiment can be divided into three sections.

In the first segment, the particle is generated and travels freely on the molecular beam axis. Subsequently, the aggregate interacts with the inhomogeneous electric field, while being between the electrodes of length l_1. Finally, the cluster travels freely over the distance l_2 until it is detected as described in Sect. 2.5. To account for the influence of the electric field on the particle's trajectory, some approximations need to be introduced. Until the particle reaches the electrodes, it travels with the velocity $v_i = v_{x,i}$ in the direction of the molecular beam axis x. All other velocity components are assumed to be zero.[1] It is important to note that the coordinate system with the axes $\{x, y, z\}$, is the space-fixed reference frame. At a later stage another body-fixed coordinate system and its connection to the space-fixed frame will be introduced. In the electric field the energy of the cluster ϵ_i consists of the internal energy $\epsilon_i^{(0)}$, i.e. rotational, vibrational and electronic contributions, and the interaction energy ϵ_i^{Int} which depends on the electric dipole moment

$$\boldsymbol{\mu} = \boldsymbol{\mu}_0 + \boldsymbol{\mu}_{\text{ind}} = \boldsymbol{\mu}_0 + \hat{\alpha}\mathbf{E} + \ldots \tag{3.1}$$

and the electric field $\mathbf{E}$. The permanent electric dipole moment $\boldsymbol{\mu}_0$ originates from the non-uniform, non-centrosymmetric charge distribution within the cluster (without applied electric field). The second term describes the field induced deformation of the charge density resulting in an additional contribution to the dipole moment. This quantity is called the induced dipole moment $\boldsymbol{\mu}_{\text{ind}}$. In general, the expression Eq. (3.1) can viewed as a Taylor expansion of $\boldsymbol{\mu}$ as function of $\mathbf{E}$ near $\mathbf{E} = 0$. Therefore, the induced dipole moment includes all orders of field induced charge density deformations. In beam deflection experiments the field strength is typically $\sim 10^7$ V/m (Sect. 2.4) and $|\boldsymbol{\mu}_0|$ is of the order of 1 D or $3.33 \cdot 10^{-30}$ Cm. The contribution of the first term of $\boldsymbol{\mu}_{\text{ind}}$ (using the typical values for α of 50–200 Å^3) compared to the permanent dipole moment is typically $(2-6)$ % whereas the other summands are at least three orders of magnitude smaller. Consequently, only the term linear in $\mathbf{E}$ is important and all other terms $(O(\mathbf{E}^2))$ can safely be neglected. For a discussion of the higher order contributions to $\boldsymbol{\mu}_{\text{ind}}$ see for example [1, 2]. The factor $\hat{\alpha}$ in the expression Eq. (3.1) is called the polarizability and is a tensor of second rank [1, 3]. Again we want to make use of the experimental conditions in order to simplify the

---

[1] This is a very good approximation for collimated clusters obtained from hard supersonic expansions. In particular, for our experimental setup v_x is typically $\sim$1200 m/s at 300 K. After collimation to 250 μm the clusters expand freely for $\sim$2.5 m and a typical full width at half maximum of 1.5 mm is measured (see molecular beam profiles). This corresponds to v_y and v_z on the order of ≈ 1 m/s.

discussion. The $\hat{\alpha}$-tensor is a symmetric tensor whose trace is much larger than the off-diagonal elements[2] [3]. Hence, the experiment is only sensitive to the diagonal elements, which additionally average during the experiment due to the fast rotational motion. As a consequence, an averaged scalar quantity $\alpha = (\alpha_{xx} + \alpha_{yy} + \alpha_{zz})/3$ is probed in experiment, which will be called polarizability from here on unless otherwise stated. If we further take into account that the "two-wire" field is aligned along the z-axis the expression for the energy

$$\epsilon_i = \epsilon_i^{(0)} - \int_0^{E_z} \boldsymbol{\mu} \cdot E_z' dE_z' = \epsilon_i^{(0)} + \epsilon_i^{\mathrm{Int}} = \epsilon_i^{(0)} - \mu_{z,i} E_z - \frac{\alpha}{2} E_z^2 \tag{3.2}$$

including the interaction contribution $\epsilon_i^{\mathrm{Int}}$ between the ith cluster and electric field can easily be derived. The quantity $\mu_{z,i}$ is the (mean) z-component of the dipole moment of particle i. Its importance and significance will become clear in the following. Moreover, in an inhomogeneous "two-wire" field the energy ϵ_i is not only a function of the electric field strength but additionally of the the particle's z-coordinate. As a result, a cluster i will experience a force

$$F_{z,i} = -\frac{\partial \epsilon_i}{\partial z} = -\frac{\partial \epsilon_i^{\mathrm{Int}}}{\partial E_z} \frac{\partial E_z}{\partial z} \tag{3.3}$$

which depends on the field gradient $\partial E_z / \partial z$ and more importantly the Stark effect $\partial \epsilon_i^{\mathrm{Int}} / \partial E_z$ [4]. Due to the force acting on the particle in the inhomogeneous field, a deflection is observed which is connected to the field induced energy change and, consequently, to $\boldsymbol{\mu}$. To quantify this qualitative statement the overall deflection d_i in z-direction needs to be known. For the "two-wire" field configuration, E_z and $E_z \cdot \partial E_z / \partial z$ are fairly constant in a small area around a given z (Sect. 2.4). Therefore, a cluster in a properly aligned beam will experience a constant force F_i. The deflection $\Delta_{i,1}$ of the cluster with mass m_i during the period $\tau_{i,1}$ within the electric field is given by $\Delta_{i,1} = F_i / m_i \cdot \tau_{i,1}^2 / 2$. Behind the electric field the momentum in z is not longer zero but is given by $F_i \tau_{i,1}$. Hence, the deflection in the field-free region before detection is described by $\Delta_{i,2} = F_i / m_i \cdot \tau_{i,1} \tau_{i,2}$ where $\tau_{i,2}$ is the period of field-free flight. As a result of this two contributions and taking the relations $v_i = l_1 / \tau_{i,1} = l_2 / \tau_{i,2}$ and Eq. (3.3) into account, the overall deflection

$$d_i = \left(\frac{\tau_{i,1}^2}{2} + \tau_{i,1} \tau_{i,2} \right) \frac{F_i}{m_i} = -\frac{(l_1^2/2 + l_1 l_2)}{m_i v_i^2} \frac{\partial E_z}{\partial z} \frac{\partial \epsilon_i^{\mathrm{Int}}}{\partial E_z} \tag{3.4}$$

can be derived. Hence, d_i depends on some experimental parameters, the particle mass as well as velocity and the Stark effect. The apparatus specific quantities can be grouped in an apparatus constant σ and calibrated by using well characterized polarizability values [5, 6]. Introducing σ and using Eq. 3.2 in Eq. 3.4 the deflection

[2] At least for atoms and non-linear molecules.

$$d_i = -\frac{\sigma}{m_i v_i^2}\left(\frac{\partial \epsilon_i^{\mathrm{Int}}}{\partial E_z}\right) = \frac{\sigma}{m_i v_i^2}(\mu_{z,i} + \alpha E_z) \tag{3.5}$$

can directly be related to the dielectric properties.

In experiment not a single particle is probed but an ensemble of particles is investigated. Consequently, we need to connect Eq. 3.5 to the experimentally measurable quantity. In our case, this is the molecular beam profile ψ (Sect. 2.1). First, we need to know how to determine m_i and v_i. Since every beam deflection experiment is coupled to a mass spectrometer, the cluster mass is inferred from the mass spectrum. Consequently, particles with identical mass are analyzed and the index i for m can be omitted. For v_i the most general approach is to measure the velocity distribution of the particles in the molecular beam. However, in supersonic molecular beams or by using velocity selectors this velocity distribution is very narrow [3] and a very good approximation is to replace v_i (in Eq. 3.5) by the mean velocity v. This mean value v is accessible by standard molecular beam techniques [5, 8, 9].

At this point we have introduced all necessary quantities and approximations in order to connect d_i with ψ_1. Due to the presence of the electric field the original beam profile $\psi_0(z)$ is shifted by d_i. For a hypothetical beam of identical i-particles the molecular beam profile with applied electric field will be given by $\psi_{i,1}(z) = \psi_0(z - d_i)$. In the case of an ensemble, the molecular beam profile with electric field

$$\psi_1(z) = \sum_i \rho_i \psi_{i,1}(z) = \sum_i \rho_i \psi_0(z - d_i) \tag{3.6}$$

is a result of the weighted summation over all ensemble members. Here, ρ_i is the deflection distribution function, describing the probability of finding the deflection d_i. When Eq. 3.6 is viewed as a convolution of the shifted ψ_0-profiles with ρ_i, it becomes apparent that all information about the dielectric response of the cluster is contained in ρ_i and the corresponding d_i values. Therefore, we need to develop a scheme that theoretically predicts ρ_i and d_i, in order to fully understand electric beam deflection experiments. Another easy transformation shows, that by using Eq. 3.5, each of the d_i values can be connected with a corresponding $\mu_{z,i}$ value. In the same way ρ_i and $\rho(\mu_{z,i})$ are easily interconvertible. Thus, the derived quantity $\rho(\mu_{z,i})$ will be called dipole moment distribution function and its theoretical predictions will be the task of the rest of this chapter.

In order to emphasize this important result, a schematic representation of the connection between $\rho(\mu_{z,i})$ and ψ_1 is shown in Fig. 3.1. A description of the beam profile ψ_1 is possible if the beam profile without electric field ψ_0 is measured experimentally and convoluted with a predicted dipole distribution function (see for example the smoothed distribution function in Fig. 3.1). If the dielectric properties used to calculate $\rho(\mu_{z,i})$ adequately describe the properties of the investigated cluster the simulated and measured ψ_1-profiles will coincide. In principle, it is possible to directly

[3] An in-depth discussions on the influence of the velocity distribution on the experimental results can be found in [7, 8].

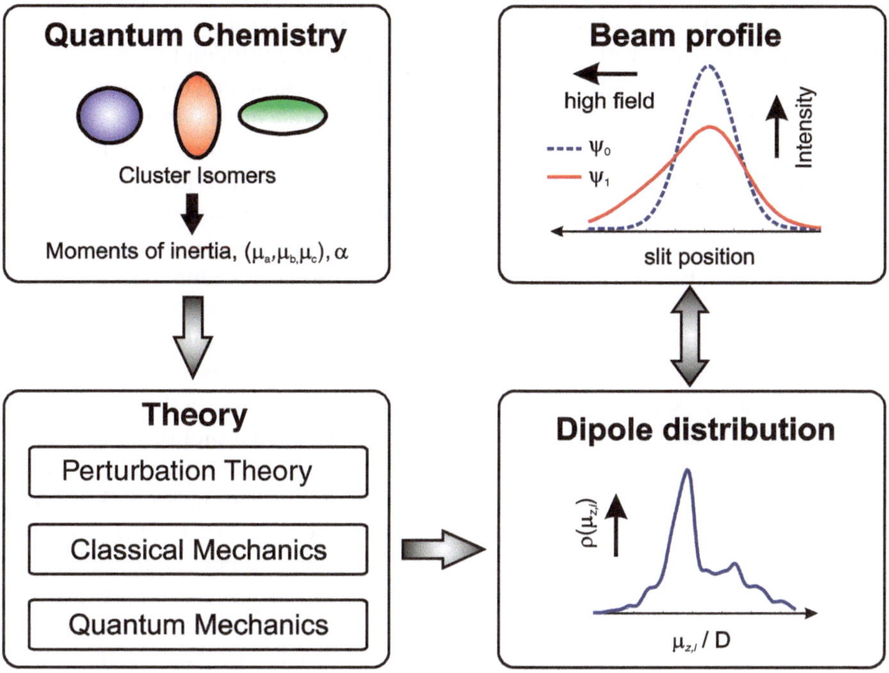

Fig. 3.1 Interpretation scheme of electric beam deflection experiments for the case of a rigid rotor. (*upper left*) An atomistic model of the clusters is developed by identifying candidate isomers employing quantum chemical methods (see Sect. 3.2). The moments of inertia, permanent dipole moment components in the molecular coordinate system (see Sect. 3.4) and polarizabilities for these isomers are inferred from these calculations. (*lower left*) These parameters are used to predict beam deflection profiles employing different methods like perturbation theory, classical or quantum mechanical simulations which are extensively discussed in Sects. 3.3–3.5, respectively. (*lower right*) As a result the dipole distribution function $\rho(\mu_{z,i})$ is obtained. (*upper right*) The beam profile with applied electric field $\psi_1(z)$ is calculated using $\psi_0(z)$, d_i and $\rho(\mu_{z,i})$ according to Eqs. 3.5 and 3.6

calculate $\rho(\mu_{z,i})$ from experiment by deconvolution and compare the predicted and extracted distribution function. Consequently, as shown in Fig. 3.1 the connection between $\rho(\mu_{z,i})$ and ψ_1 is reversible. Until now this has only been achieved for magnetic deflection experiments and is mentioned here for completeness [10]. So far, we have discussed the basic experimental and theoretical quantities, in order to calculate $\rho(\mu_{z,i})$. To give an idea how to compute ρ_i a first step is to discuss the variables that the distribution function depends on. First, we will assume that all clusters are in the same electronic state and only the rovibrational degrees of freedom will be considered explicitly. Therefore, in general ρ_i will be a function of the energy stored in the rotational and vibrational degrees of freedom characterized by T_{rot} and T_{vib} called the rotational and vibrational temperature,respectively. In order to describe the rotational and vibrational motion and their influence on ρ_i further quantities that characterize these motions, i.e. the principle moments of inertia **I** and the harmonic oscillator frequencies ν_0, are needed. Second, ρ_i depends on $\boldsymbol{\mu}_0$, α and E_z (Eq. 3.5). This tremendous number of variables illustrate, that it is a difficult task to predict ρ_i from first principles. In particular, excited vibrations and interactions between the rotational and vibrational motion are difficult to incorporate in a model (see Sect. 3.6). Fortunately, in experiment the clusters can be cooled by cryogenic supersonic expansions (see Chap. 2) resulting in a nearly rigid particle skeleton. Within this so called rigid rotor assumption, the problem of predicting $\rho(\mu_{z,i})$ is simplified and only the rotational motion of a cluster in an electric field needs to be considered. Depicted in Fig. 3.1 are the various methods that can be used to achieve this task for a rigid cluster. However, if the particles are not longer rigid and undergo vibrations and/or isomerizations the presented methodology breaks down. Therefore, the description of molecular beam profiles for these "floppy" clusters will be presented separately in Sect. 3.6 and all what follows is only valid for rigid clusters. In the scenario of a rigid rotor, the influence of the electric field on the rotational motion of the cluster can either be modeled by using perturbation theory, classical or quantum mechanics (see Fig. 3.1). The different methods will be discussed in detail in Sects. 3.3–3.5. In order to perform the simulations, there is only one ingredient missing (Fig. 3.1). This is an atomistic model of the cluster and its corresponding properties. It is a separate problem to correctly predict energetically low-lying cluster structures and have adequate $\boldsymbol{\mu}_0$, α and $\mathbf{I}$ values available to simulate the electric beam deflection behavior. Therefore, the following Sect. 3.2 briefly deals with state-of-the-art methods used to locate cluster structures and introduces the essentials to extract the needed set of parameters from quantum chemical computations.

3.2 Quantum Chemical Prediction of Cluster Structures and Dielectric Properties

3.2.1 Predicting Cluster Structures Using Global Optimization Techniques

The prerequisite for every beam deflection simulation that aims to investigate the structures of gas phase clusters, is to deduce the parameters $\boldsymbol{\mu}_0$, α and $\mathbf{I}$ from quantum chemical calculations. However, what is the structure of the corresponding gas-phase species? This is one of the most essential questions in cluster physics and the answer is not trivial. The simplest idea is to guess the structures of the clusters based on the structural motifs of molecules or solid-state compounds. This approach of educated guessing can lead to erroneously wrong ground state (GS) structure predictions. One of the best documented examples is the structure of the fullerene C_{60}. Before the discovery of C_{60} the modifications graphite and diamond were known [11]. Based on these modifications the structure of C_{60} should either be a truncated graphite-sheet, only containing hexagons, or a dense nearly spherical aggregate with sp^3 coordinated atoms like in diamond. Today it is well known that C_{60} is a hollow structure consisting

12 pentagons and 20 hexagons [12]. Another idea is to create all possible structures by hand. While this approach is widely used for very small particles, the number of possible isomers roughly increases exponentially with the number of atoms N, making this procedure intractable for larger clusters [13, 14]. The situation is further complicated for multi-metallic particles, so called nanoalloys. Beside the structural isomers a huge number of homotops, describing the permutational isomerism, for every nanoalloy cluster structure exist [13, 15].

Hence, a rational approach must be used in order to systematically locate possible isomers. In the recent years, several so called global optimization routines were developed. While, all of the methods employ different searching algorithms the overall goal is to locate the global minimum (GM) or GS structure on the potential energy surface (PES) for a given size and composition. The two most widely used methods are called basin-hopping (BH) and genetic algorithm (GA) [14]. The principle of these two routines are depicted in Fig. 3.2. First, the PES is modeled by some interatomic potential (black line). We will come back to the issue of a proper modeling of the PES at a later stage. In general, the energy is a function of $3 \cdot N$ Cartesian coordinates $\mathbf{X}$. For simplicity, the PES in Fig. 3.2 is shown in dependence of the generalized coordinates $\{\mathbf{X}\}$. Each of the minima is assigned to a corresponding structure. Some of these structures are schematically shown as ellipses and circles. How the PES can be systematically explored locating the lowest energy isomers? In the BH approach, introduced by Li and Scheraga [16] and formulated more strictly by Wales and Doyle [17], this is done by creating a trial structure (randomly or by hand) and in a first step this structure is energetically optimized. Hence, the search starts from one of many possible minima. The main idea of this method is to randomly change the coordinates of one or several atoms of the initial cluster within a predefined trust radius. This avoids dissociation or vaporization of the cluster. The newly generated structure is again energetically relaxed. In the case, that the geometrical change was sufficient to at least overcome one of the transition states surrounding the initial structure, the new structure will correspond to another isomer, located in a nearby minimum. Due to the local geometry optimization not the PES (Fig. 3.2, solid black line) is studied but it is the transformed PES that is explored (Fig. 3.2, broken black line) [17]. The energy difference $\Delta\epsilon$ between the new and old isomer are compared by the Metropolis Monto-Carlo (MC) criterion, in order to decide if the new structure is accepted [18]. The random BH-move is accepted, if $\Delta\epsilon < 0$ or for $\Delta\epsilon > 0$ if $\exp[-\Delta\epsilon/(k_\mathrm{B}T)]$ is larger than a randomly generated number. Here k_B is the Boltzmann constant and T is the simulation temperature. Therefore, this procedure is equivalent to hopping (see Fig. 3.2) from one minimum to another. However, an intrinsic problem of the BH algorithm is that the code tries to move between minima by random changes of the geometry but, if a local basin is considerably lower in energy than all surrounding minima and is surrounded by high transition state barriers, it is very unlikely that the search can escape this minimum and, hence, it is trapped. Recent research efforts resulted in strategies which can overcome this problem [19, 20]. A simple example is the so called jumping move (see Fig. 3.2) [19]. In the case the structure of the particle does not change for several MC steps the temperature is raised to ∞, equivalent to always accepting this move,

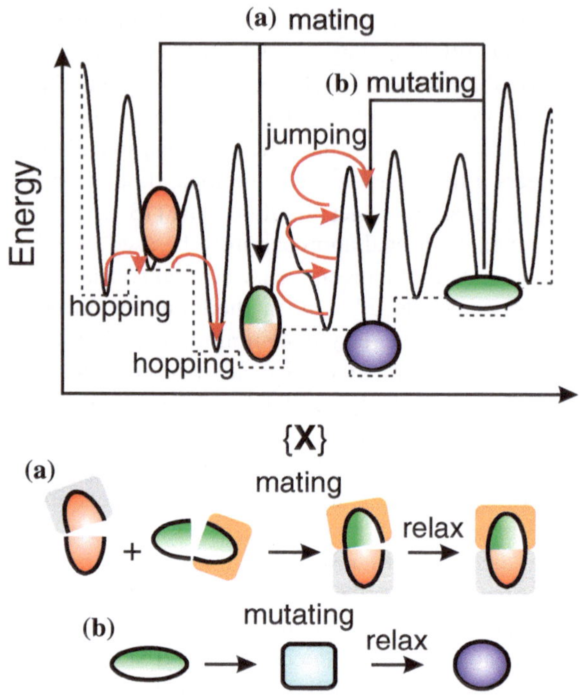

Fig. 3.2 Illustration of the basic principles of the BH and GA global optimization strategies. The PES (*solid line*) is shown as a function of the generalized coordinates {**X**}. For some minima the corresponding cluster structures are represented as *ellipses* and *circles*. Since the energy is optimized in every global optimization step, it is the transformed PES (*broken line*) that is searched. The basic searching strategies of the BH and GA are included in the figure and are highlighted by *arrows* and *letters*, respectively. **a, b** Graphical representation of mating and mutation. The effect of these operations is shown in the *upper part* of the figure

and the structure is changed several times without further geometry optimization. This allows to jump out of a deep basin. After this jumping move the BH routine continuous with normal hoping steps. The energetically lowest structure located after finishing the predefined number of MC steps is the putative GM.

A somewhat different approach is used in GA searches. The GA uses evolutionary principles in order to locate the GM [13]. The cluster structure is represented by a certain set of coordinates or variables defined as genes. The task is to globally optimize the values (alleles) for the complete set of genes (forming a chromosome). Contrary to the BH search, not a single starting structure is used but a number of isomers, called individuals, are grouped to form the initial population. Each structure is energetically optimized and the fitness value $f_i(\eta_i)$ of the individual i with energy ϵ_i depending on the relative energy scale

$$\eta_i = \frac{\epsilon_i - \epsilon_{\min}}{\epsilon_{\max} - \epsilon_{\min}} \tag{3.7}$$

is assigned to each structure. Here $\epsilon_{\min}$ and $\epsilon_{\max}$ is the lowest and highest energy value in the population, respectively.[4] A systematic energetic improvement of the cluster structure is possible if different genetic operators are applied to the cluster chromosome. There are several ways of deciding which cluster should be used in a

[4] For possible functional forms of $f_i(\eta_i)$ see for example Ref. [13].

genetic operation. Most of these routines are based on $f_i(\eta_i)$ [13]. The higher the fitness value, the more likely the structure is used in a genetic operation. The two most important genetic operations are depicted in Fig. 3.2. For mating or crossover, two individuals of the population are chosen. In the simplest form of the so called "cut-and-splice" operation introduced by Deaven and Ho [21], each cluster is cut at a single random position (and orientations) and the two fragments are spliced with random orientations, leaving the total number of atoms and the composition constant (see Fig. 3.2a). In this way, energetically beneficial fragments of the parent clusters are transfered to an offspring isomer. Another genetic operation, which is shown in Fig. 3.2b, is the mutation. The chromosome of the individual is changed randomly, i.e. the translation of a single atom, the rotation of fragments, the interchange of atom types or the generation of a completely new structure. Subsequent to these operations offspring and mutants are locally optimized with respect to the energy. Therefore, it is the transformed PES (see Fig. 3.2, broken black line) that is investigated in most GA searches. Now, the total number of individuals exceeds the predefined population size. In order to decrease the number of clusters to the allowed population size, some isomers are deleted based on their fitness-value and structural motif [22]. This process is a natural selection step and is inspired by selection strategies found in nature. All the described processes of fitness evaluation, genetic operations, natural selection and relaxation are defined to form a generation. All processes are repeated until a predefined convergence criterion, for example the energy of the lowest lying isomer, does not change for a certain number of generations. The lowest isomer in the last generation is predicted to be the putative GM. The herein presented description of the BH and GA approach is only intended to briefly introduce the basic working principles of some global optimization strategies. For a more detailed description of the BH and GA methods and possible variants, the reader is referred to Refs. [13, 19, 20, 22–24].

As shown above, in principle it is possible to use one of the mentioned (or another) global optimization techniques to systematically locate the GM or energetically low-lying isomers. However, an interatomic potential model is needed to describe the bonding within the cluster and, hence, to model the corresponding PES. The simplest way is to use some kind of model or empirical potential. For larger metallic clusters ($N \geq 30$) the Gupta [25] or Sutton-Chen [26] potential is used to perform these kind of global optimizations. It is expected that clusters of this size behave very similar to the bulk, justifying the use of these potential energy functions. By decreasing the total number of atoms it will become possible to use tight-binding approaches [27] or *ab initio* calculations [28–30] in order to study the PES. This is particularly important for clusters for which the chemical bonding massively differs from the bulk. No matter what method is used to describe and search the PES, all will lead to putative GM and energetically low-lying isomers. Certainly, the more realistic the modeling of the interatomic bonding, the better these structures will correlate with the isomers studied in experiments. How to extract the required properties from the theoretically predicted isomers, will be discussed in the next paragraph.

3.2.2 Quantum Chemical Predictions of the Dielectric Properties

At this point, we come back to the original problem of computing α, $\boldsymbol{\mu}_0$ and **I**. This can be done for the candidate structures located with one of the above mentioned methods. In order to calculate all three necessary properties, the best way is to use an *ab initio* method which will give the best possible description of the geometric and electronic structure. Hence, structures found by global optimization routines are in a next step used as starting structures for Hartree-Fock (HF) [31], other wave function based [31–33] or density functional theory (DFT) [34, 35] calculations [36, 37].[5] A good general overview on how to calculate various properties using quantum chemical methods is given in Ref. [38].

For wavefunction based methods the problem is to solve the time-independent Schrödinger equation $\hat{H}\Psi = \epsilon\Psi$ taking the exact spin-free, non-relativistic Hamiltonian

$$\hat{H} = -\frac{\hbar^2}{2m_e}\sum_j \nabla_j^2 - \sum_j\sum_M \frac{Z_M e^2}{4\pi\epsilon_0 \mathbf{r}_{jM}} + \sum_j\sum_{k>j} \frac{e^2}{4\pi\epsilon_0 \mathbf{r}_{jk}} + \sum_L\sum_{M>L} \frac{Z_L Z_M e^2}{4\pi\epsilon_0 \mathbf{R}_{LM}} \tag{3.8}$$

within the Born-Oppenheimer approximation [39] for n interacting electrons in an N atom molecule without and with the presence of an external field **E** into account. For convenience, Eq. 3.8 is given without incorporating the influence of **E**. The exact electronic ground state energy is given by ϵ, ∇ is the Nabla operator, Ψ is the n-electron ground state wavefunction, m_e, e, $\hbar$ and ϵ_0 are the electron mass, unit charge, the Planck constant h divided by 2π and the permittivity in vacuum, respectively. The corresponding atomic number is represented by Z_M, the electron-nuclei, electron-electron and nuclei-nuclei separation are given by $\mathbf{r}_{jM}$, $\mathbf{r}_{jk}$ and $\mathbf{R}_{LM}$, respectively. Since this equation cannot be solved exactly for more than one electron due to the presence of the electron-electron interaction term, quantum chemists have introduced various approximations to solve this problem numerically. The most basic approximation that conserves the antisymmetric character of the fermionic wavefunction is called HF method. Within this approximation it is assumed that the n-particle wavefunction can be written as an anti-symmetric product of one-electron functions, i.e. orbitals. The resulting wavefunction

$$\Psi^{\mathrm{HF}} = \frac{1}{\sqrt{n!}} \begin{vmatrix} \chi_j(\mathbf{r}_1) & \chi_k(\mathbf{r}_1) & \cdots & \chi_l(\mathbf{r}_1) \\ \chi_j(\mathbf{r}_2) & \chi_k(\mathbf{r}_2) & \cdots & \chi_l(\mathbf{r}_2) \\ \vdots & \vdots & & \vdots \\ \chi_j(\mathbf{r}_n) & \chi_k(\mathbf{r}_n) & \cdots & \chi_l(\mathbf{r}_n) \end{vmatrix} \tag{3.9}$$

[5] Some of the properties can be deduced from simple model calculations which maybe give an reasonable agreement with experimental findings. Here, we will concentrate on the most generally applicable approach using *ab initio* methods. For the prediction of α by simple model calculations see Refs. [3, 34].

called Slater determinant consists of the spin-orbitals χ_j.[6] Since the choice of χ_j is not unique, the basic idea is to chose a set of spin-orbitals and optimize these orbitals by minimizing the HF energy ϵ^{HF} according to the variational principle. This will result in a set of effective one-electron eigenvalue equations

$$\hat{f}(\mathbf{r})\chi_j(\mathbf{r}) = \epsilon_j \chi_j(\mathbf{r}) \tag{3.10}$$

with the orbital energy ϵ_j and the Fock operator defined as

$$\hat{f}_j = -\frac{\hbar^2}{2m_e}\nabla_j^2 - \sum_M \frac{Z_M e^2}{4\pi\epsilon_0 \mathbf{r}_{jM}} + \mathrm{v}_j^{\mathrm{HF}} \tag{3.11}$$

which are solved self-consistently. In Eq. 3.11, $\mathrm{v}_j^{\mathrm{HF}}$ is the averaged effective electrostatic potential experienced by electron j, originating from all other electrons. It is known that there are two possibilities for the spin-state of each electron but how can the spatial part of ϕ_j be described? For atoms, ϕ_j is expanded as a superposition of known functions called basis functions, whereas for molecular systems the so called molecular orbitals (MOs) are used which are chosen to be a linear combination of atomic orbitals (LCAO) [31]. The variational freedom is given by the expansion coefficients used in the basis set. Since the basis set will only be finite in real computations, the size of the basis set may introduce a source of error.[7] After having solved the above outlined HF problem for a given set of nuclear coordinates the atomic positions are changed and the HF procedure is repeated until the change of the electronic and nuclear energy, i. e. the force acting on each atom, is smaller than a predefined convergence criterion.[8] For this step, standard minimization techniques discussed in Ref. [40] are used. After this whole procedure we have found the optimal geometry and wavefunction using the HF approximation for a given basis set expansion without electric field. Certainly, the HF method is an approximation and, in particular, the effective potential and the use of a single Slater determinant do not account for all of the exchange and correlation effects. A systematic improvement of the HF wavefunction and energy is achieved by Møller-Plesset (MP) perturbation theory [31, 41], coupled cluster (CC) calculations [33] or multi-determinant approaches [32]. We do not intend to introduce all of these methods since the basic principles of calculating the required parameters will be very similar, even if the implementation and the actual numerical calculations can be tremendously more difficult. However, we must keep in mind that only the parameters extracted from advanced quantum methods will give acceptable values for α, $\boldsymbol{\mu}_0$ and $\mathbf{I}$ (see Table. 3.1). The actual level of theory that is needed depends on the system under investigation. Additionally, relativistic effects

[6] All following considerations will only be valid for closed shell atoms and molecules. The procedure is very similar for open-shell systems but the resulting equations are somewhat more complicated. For open shell systems the reader is referred to Ref. [31].

[7] The larger the basis the smaller will be ϵ^{HF}. If ϵ^{HF} does not change when the basis set is increased the HF limit is reached, i.e. the best result the HF approximation can offer. In a similar way, predictions from other quantum chemical methods can be improved.

[8] This is only of importance for molecular systems.

Table 3.1 Comparison of theoretical and experimental α and μ_0 values in Å^3 and D, respectively, for various atomic, molecular and cluster systems

	α_{theo}/Å^3	α_{exp}/Å^3		$\mu_{0,\text{theo}}$/D	$\mu_{0,\text{exp}}$/ D
He/QED[a]	0.21[j]	0.21 ± 0.01 %[j]	LiF/MP4[f]	6.25	6.28 ± 0.01[j]
Na/CCSDT[b]	24.14[j]	24.11 ± 0.12	3-Aminophenol/CASPT2[g]	2.16	2.33 ± 0.01[j]
Ba_2/HF[c]	111.53	103.20 ± 10 %	3-Aminophenol/B3LYP[g]	2.49	2.33 ± 0.01[j]
Ba_2/CCSD(T)[c]	97.88	103.20 ± 10 %	Na_{14}/B3LYP[h]	1.17	0.02 ± 0.02
Na_{20}/PW91LDA[d]	293.20	304.4 ± 7.60	Na_{14}/vdW−DFT[h]	0.03	0.02 ± 0.02
Anthracene/B3LYP[e]	26.29	25.93 ± 10 %	Pb_{18}/MP2[i]	0.40	0.59 ± 0.11

[a] Quantum electrodynamic (QED) calculations and dielectric constant measurements [53, 54]
[b] CCSDT (coupled cluster with single double and triple excitations) and interferometry results [55, 56]
[c] Scalar relativistic calculations and beam deflection results [5]
[d] DFT calculations and beam deflection measurements [57, 58]
[e] DFT and Stark-modulated laser spectroscopy results [59, 60]
[f] MP4 theory and molecular beam electric resonance results [61, 62]
[g] Stark-modulated rotational spectroscopy and the corresponding CASPT2 (second order perturbation complete active space method) and B3LYP calculations for the *cis* isomer [63]
[h] B3LYP and van-der-Waals (vdW) corrected DFT calculations and beam deflection measurements [58, 64, 65]
[i] MP2 calculations and beam deflection results [66, 67]
[j] This value is more precise than stated in the table

may play a crucial role for these properties [42]. Regardless of these limitations we can assume that the computed geometry and wavefunction using the HF approach gives a first approximation to the exact value of the molecular properties. Improved results maybe further obtained from sophisticated wave function based methods.

The quantity that is easily extracted from the calculations, is the moment of inertia tensor

$$\hat{I} = \begin{pmatrix} \sum_M m_M(y_M^2 + z_M^2) & -\sum_M m_M x_M y_M & -\sum_M m_M x_M y_M \\ -\sum_M m_M x_M y_M & \sum_M m_M(x_M^2 + z_M^2) & -\sum_M m_M y_M z_M \\ -\sum_M m_M x_M z_M & -\sum_M m_M y_M z_M & \sum_M m_M(x_M^2 + y_M^2) \end{pmatrix}, \tag{3.12}$$

that is, after transformation in the center of mass coordinate system, deduced from the final cluster geometry. Here the sum runs over all atoms N, m_M is the mass of the corresponding nucleus and the set $\{x_M, y_M, z_M\}$ are the coordinates of all atoms. In the next step we transform $\hat{I}$ into the principle coordinate system giving a diagonal matrix. This new matrix is in short written $\mathbf{I} = (I_a, I_b, I_c)$ and only contains the principle moments of inertia.[9] The coordinate system corresponding to this principle axis system defines the body-fixed coordinate system. Its axis are given by $\{a, b, c\}$ and all molecular properties will be given in this coordinate system.

[9] For all what follows we define the principle moments of inertia to fulfill the relation $I_a \geq I_b \geq I_c$. Please note that this different from the definition most commonly used. In terms of moments of inertia a spherical rotor will have $I_a = I_b = I_c$. Breaking this symmetry will lead to a prolate ($I_a = I_b > I_c$) and an oblate ($I_a > I_b = I_c$) rotor which both are classified to be symmetric tops. Note that the symmetry axis of a prolate top is c and for an oblate rotor it is a. For an asymmetric rotor the moments of inertia fulfill the relations $I_a \neq I_b \neq I_c$.

The next task is to calculate the polarizability. In order to computationally extract this property from quantum chemical considerations, Eq. (3.2) can be rewritten and integrated with respect to the electric field, giving

$$\epsilon = \epsilon^{(0)} - \int_0^{\mathbf{E}} \boldsymbol{\mu} \cdot \mathbf{E}' d\mathbf{E}' = \epsilon^{(0)} - \left.\frac{\partial \epsilon}{\partial \mathbf{E}}\right|_{\mathbf{E}=0} \mathbf{E} - \frac{1}{2} \left.\frac{\partial^2 \epsilon}{\partial \mathbf{E}^2}\right|_{\mathbf{E}=0} \mathbf{E}^2 + \ldots. \quad (3.13)$$

This illustrates, by comparison with Eq. (3.2), that $\hat{\alpha}$ can be connected with $-\partial^2\epsilon/\partial\mathbf{E}^2|_{\mathbf{E}=0}$ (including all possible derivatives with respect to the electric field). Consequently, the task is to find the second-order energy derivatives with respect to the electric field, in order to calculate $\hat{\alpha}$ and therefrom α. The most straightforward way is to include the operator $\hat{H}' = -e\mathbf{E}\sum_j \mathbf{r}_j$ ($\mathbf{r}_j$ indicates the coordinates of electron j) in the quantum chemical calculations and repeating the HF calculations for a set of finite field strength. By numerical differentiation the corresponding derivatives can easily be calculated [43]. Two more commonly used methods are uncoupled and coupled perturbed calculations. For these calculations $\hat{H}'$ is interpreted as perturbation operator and is used within the HF procedure. By expanding the corresponding set of wavefunctions in the form $\Psi = \Psi^{(0)}\Psi^{(1)}$ for the unperturbed and $\Psi = \Psi^{(0)} + \Psi^{(1)} + \ldots$ for the perturbed approach, where $\Psi^{(0)}$ is equivalent to Eq. (3.9) and $\Psi^{(1)}$ is the first perturbation correction of $\Psi^{(0)}$, a set of uncoupled and coupled equations are obtained, respectively. These are solved to obtain the energy derivatives with respect to the electric field and, hence, $\hat{\alpha}$. For a review on the development of these methods see Ref. [44]. On the one hand, the unperturbed method is easier to solve but on the other hand, present days computations can readily solve the more accurate perturbed equations, which include the electron correlation effects due to the interaction between different orbitals in the presence of the electric field. For higher-order correlation methods like MP, CC or configuration interactions (CI) similar finite field and perturbation methods (see Ref. [45] for high accuracy calculations) are available while modern quantum chemical approaches use (linear) response theory calculations to evaluate the time-dependent and -independent properties. A description of the more complex linear-response calculations is beyond the scope of this manuscript but these methods have been reviewed recently [46]. The final method introduced here is the so called sum-over-states approach which is based on the second order perturbation theory expression

$$\alpha_{kl} = 2e^2 \sum_m \frac{\langle \Psi^{(0)} |k| \Psi_m\rangle \langle \Psi_m |l| \Psi^{(0)}\rangle}{\epsilon_m - \epsilon^{(0)}} \quad (3.14)$$

for the klth $\hat{\alpha}$-component [47]. The GS wavefunction $\Psi^{(0)}$ in this expression is perturbatively influenced by the excited states Ψ_m via the kth and lth electron coordinate operators. The difference of the GS energy $\epsilon^{(0)}$ and the energies of the excited states ϵ_m are inversely proportional to the polarizability component α_{kl}. Therefore, excited states which are close in energy to the GS, contribute most significantly to the polarizability.[10] Beside the qualitative understanding Eq. 3.14 can offer, quantum-chemical methods that predict ϵ_m and Ψ_m can be used to calculate the polarizability via this equation [3, 45].

The final quantity we have to predict is $\boldsymbol{\mu}_0$. From Eq. 3.13 it is obvious, that $-\boldsymbol{\mu}_0\mathbf{E}$ is a first order correction to the field free energy. As a consequence and in the spirit of perturbation theory, the dipole moment is given by $\boldsymbol{\mu}_0 = \langle \Psi^{(0)} | \hat{\boldsymbol{\mu}} | \Psi^{(0)} \rangle$. This expression is equivalent to the expectation value of the dipole moment operator $\hat{\bar{\mu}} = -e\sum_j \mathbf{r}_j + e\sum_M Z_M\mathbf{R}_M$ of the electronic ground state. The indices of the summation are identical to the definition introduced before. By using $\hat{\boldsymbol{\mu}}$ and the fact that electrons are indistinguishable the dipole moment is given by

$$\bar{\mu}_0 = \int p(\mathbf{r})\mathbf{r}d\mathbf{r} + e\sum_M Z_M\mathbf{R}_M \tag{3.15}$$

in which $p(\mathbf{r})$ represents the electron density. The dipole moment is directly inferred from the GS wavefunction and geometry without further computations. We have to stress that the computed value of $\boldsymbol{\mu}_0$ is unique. This is in contrast to the problems arising when calculating partial charges [31]. Hence, dipole moments calculated from Eq. 3.15 and from partial charges are in most cases different. The discussion above was based on methods using a n-electron wavefunction. The advantage of these methods is that a systematic procedure exists to improve the calculation results. This is done by systematically improving the description of exchange-correlation effects and by increasing the basis set size. Unfortunately, this procedure becomes computationally very expansive and is only used for very small or light cluster systems.

Therefore, at the end of this Section we briefly want to introduce another very widely used technique called DFT [34, 35]. In DFT calculations not the time-independent Schrödinger equation or the ground state wavefunction is used, but the energy is expressed as a functional of the electron density

$$\begin{aligned}\epsilon[p(\mathbf{r})] = T[p(\mathbf{r})] + \frac{1}{2}\int \frac{p(\mathbf{r})p(\mathbf{r}')}{4\pi\epsilon_0(|\mathbf{r}-\mathbf{r}'|)}d\mathbf{r}d\mathbf{r}' \\ + \int p(\mathbf{r})\left[v_{\mathrm{i}}(\mathbf{r}) + v_{\mathrm{ext}}(\mathbf{r})\right]d\mathbf{r} + \epsilon_{\mathrm{xc}}[p(\mathbf{r})]\end{aligned} \tag{3.16}$$

depending on the kinetic energy functional $T[p]$, the electron-electron repulsion term, the ionic and possible external potential v_{i} and v_{ext} as well as the exchange correlation functional $\epsilon_{\mathrm{xc}}[p]$. It was shown by Hohnberg and Kohn that $p(\mathbf{r})$ is fully sufficient to describe all desired properties of the electronic ground state and that the best result is found by variationaly changing $p(\mathbf{r})$ in order to minimize the energy $\epsilon[p]$ [48]. There are various approximations to Eq. 3.16 [3, 34] but at present the obtained results are not accurate enough. Hence Eq. 3.16 is mainly used to perform model calculations. A practicable solution is found by expanding the electron density in the finite basis of well behaved functions $\{\phi_j\}$. By adopting ideas of HF computations, $\epsilon[p(\mathbf{r})]$ is minimized by optimizing $\{\phi_j\}$ and a set of single-particle equations of the

[10] For a system with only two energy levels Eq. 3.14 simplifies to Eq. 1.1.

form

$$\left[-\frac{\hbar^2}{2m_e}\nabla_j^2 + \mathrm{v}_\mathrm{j}^\mathrm{KS}\right]\phi_j = \epsilon_j\phi_j \tag{3.17}$$

are obtained. These so called Kohn-Sham (KS) equations [49] can be solved self-consitently.[11] In analogy to Eq. 3.11, ϵ_j is the energy of the KS orbital ϕ_j. By comparison with Eq. 3.10, this expression is very similar to the HF result, apart form the potential $\mathrm{v}_\mathrm{j}^\mathrm{KS} = \mathrm{v}_\mathrm{0,j}^\mathrm{KS} + \mathrm{v}_\mathrm{ext} + \delta\epsilon_\mathrm{xc}[p]/\delta p(\mathbf{r})$, which consists of three summands. The first two being the effective electron repulsion term $\mathrm{v}_\mathrm{0,j}^\mathrm{KS}$ and the external potential v_ext due to the nuclei or other applied fields. Most importantly, the effective potential in KS-DFT contains the $\delta\epsilon_\mathrm{xc}[p]/\delta p(\mathbf{r})$ term, which is related to the chosen exchange-correlation functional. While we have learned that v_j^HF does not incorporate all exchange-correlation (xc) effects and can only be improved by more demanding calculations, the strength of DFT is that a proper choice of the xc-functional, in principle, allows to include all xc-effects with the computational expanse very similar to HF calculations. Unfortunately, $\epsilon_\mathrm{xc}[p]$ is not known and various approximations and semi-empirical models for this functional have been developed and used over the last decades [51, 52]. All of these functionals have their disadvantages and benefits but the biggest problem is that a general and systematic approach to improve these methods is missing.[12] Nevertheless, the benefits in computational cost compared to wavefunction based methods, which perform similarly, outrun these drawbacks, especially for larger cluster systems. After a proper experimental or theoretical calibration, DFT is a very widely and successfully used method [34, 35, 51]. Since Eqs. 3.10 and 3.17 are very similar, nearly all methods introduced above for geometry optimization, calculation of **I**, α and $\boldsymbol{\mu}_0$ exist for DFT, too. For example there are finite field [68], coupled perturbed KS [69, 70] and linear response theory calculations (mostly within time-dependent DFT) [70, 71] in order to compute α. Furthermore, the optimized DFT geometry is used to deduce **I** and the final $p(\mathbf{r})$ is employed to compute $\boldsymbol{\mu}_0$ via Eq. (3.15).

Therefore, a broad variety of quantum chemical methods exist which will yield the parameters **I**, α and $\boldsymbol{\mu}_0$. In Table 3.1 some calculation results for α and $|\boldsymbol{\mu}_0| = \mu_0$ for a selection of atomic, molecular and cluster systems are presented to highlight the accuracy and problems of present day computations of dielectric properties. The quality of predicted **I** values is, in general, satisfying and, hence, the major problem in determining the principle moments of inertia is to find the GM.[13] The situation is

[11] In cluster physics a well known approximation to these equations is the Jellium model. In this model, the valence electrons are treated explicitly and the nuclear charge that balances the valence electrons is assumed to be distributed uniformly. While this approximation is not restricted to DFT it was mainly used in this scientific community. For an in-depth review of this method see [50].

[12] A systematic improvement of the results is achieved by increasing the basis set size but so far there is no strict way to systematically account for xc effects.

[13] In order to compare to experiments like beam deflection measurements this conclusion is valid. In the case of high resolution spectroscopy techniques, highly correlated methods and large basis sets have to be used in order to obtain the accuracy needed to describe the experimental findings. See Ref. [72] for an example.

very similar for computations of α. No matter if open- or closed-shell atoms, heavy element dimers, clusters or hydrocarbons with delocalized electrons, the computational findings agree with experiments (see Table 3.1). Particularly interesting is the fact that even HF and CCSD(T) calculations for the heavy element dimer Ba_2 only differ by $\sim$10 %. Therefore, the value of α can safely be exctracted from nearly all quantum chemical calculations, resulting in a fairly good agreement with experimental observations. In terms of comparing with beam deflection measurements, for which typically uncertainties between 3 and 10 % are observed (Chap. 4), the results of the computations are adequate. A different observation is made by closely inspecting the μ_0 values of Table 3.1. For a small molecule like LiF the experimental high precission results are nearly reproduced by MP4 calculations. However, the experimental dipole moment from microwave spectroscopy investigations for the molecule *cis*-3-Aminophenol, only containing first and second row elements, is not reproduced by different computational approaches [63]. Much more problematic, as highlighted with the two values in Table 3.1, is the fact that results from different methods can differ by 10 and 15 % and so far there is no strict correlation between the quality of the theoretical method and the obtained μ_0 value. For cluster systems this discrepancy between theoretical and experimental dipole moments can even become worse as shown in Table 3.1 for Na_{14}. Though, the interpretation of beam deflection measurements, from which nearly all μ_0 values for larger clusters are extracted, is not straightforward, the difference with theory can not only be caused by experimental inaccuracies. The theoretical reason for the massive differences between the various computational methods is not clear at the moment and this subject is a topic of current research [73]. So far the theoretical methods that aimed to describe the dielectric properties were benchmarked by using polarizability values but this can be problematic as outlined above [74].

Nevertheless, we have introduced the methods required to develop an atomistic cluster model and calculate **I**, α and $\boldsymbol{\mu}_0$. In the next Sections this knowledge is used to adequately describe the rotational motion of a rigid cluster in an electric field, thereby introducing different interpretation schemes.

3.3 Rigid Rotor in a Weak Electric Field: Beam Deflection and Distribution Function

Before we start to discuss the behaviour of clusters in an electric field in much detail and describe the methodologies which allow an accurate calculation of $\rho(\mu_z)$ [or the corresponding ψ_1], we want to develop a simple physical model for the electric beam deflection and dipole moment distribution function. In this way, we will be able to identify the most important characteristics of $\rho(\mu_z)$ and the corresponding deflection profiles ψ_1. This, hopefully, will enable the reader to easily interpret most of the results discussed in the literature and assist the understanding of all following sections and chapters.

For this purpose we want to consider a rigid cluster in a weak electric field [74–76]. This means that the ratio of the interaction energy with the electric field and the energy stored in the rotational degrees of freedom is very small, i.e. $\boldsymbol{\mu}\mathbf{E}/(k_{\mathrm{B}}T_{\mathrm{rot}}) \ll 1$. Hence, the energy ϵ_i consists of the field free energy $\epsilon_i^{(0)}$ and the field interaction energy $\epsilon_i^{\mathrm{Int}}$ (see Eq. 3.2). In a weak electric field $\epsilon_i^{\mathrm{Int}}$ will be small, too. Consequently, the rotational motion will not be influenced by the applied electric field. Therefore, a first approximation is found by simply including a first order energy correction to $\epsilon_i^{(0)}$, i.e. first oder perturbation theory (FOPT) is applied so that $\epsilon_i^{\mathrm{Int}} \approx \epsilon_i^{(1)}$. To further simplify our model, we will make the unphysical assumption that the cluster is spherically symmetric. This assumption implies that the particle can not possess a permanent electric dipole moment, due to the presence of an inversion center. However, it will become clear at a later stage of the discussion, that clusters with finite μ_0 only show a very small deviation from the spherical symmetry and, hence, the model is a good starting point to analyze experimental findings.

To assess to classical FOPT energy a very simple model is shown in Fig. 3.3. The dipole moment $\boldsymbol{\mu}_0$ that is aligned along the symmetry axis of the particle, rotates around $\mathbf{J}$. In addition, the angular momentum vector $\mathbf{J}$ itself rotates around the laboratory z-axis and can, in the discussed classical limit, point in all possible directions. The influence of α is not included in the Fig. 3.3 for the sake of clarity. The effective first interaction energy of the permanent dipole moment with the electric field is obtained by projecting $\boldsymbol{\mu}_0$ on z and averaging over the rotational motion, which is fast compared to the experimental time span (Eq. 3.34). Since the symmetry axis of cluster rotates around $\mathbf{J}$ the projected dipole moment along the $\mathbf{J}$-axis is $\mu_0 \cos\beta = \mu_0 K/J$ (see Fig. 3.3). Here, $K\hbar$ is the projection of $\mathbf{J}$ along the symmetry axis of the parti-

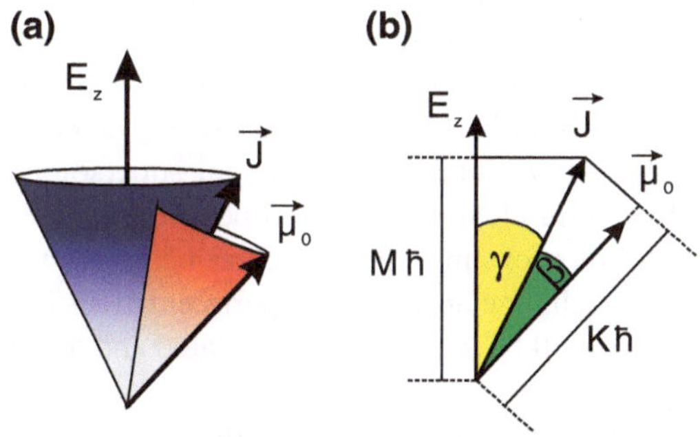

Fig. 3.3 Simple geometrical construction for a classical rigid rotor in a weak electric field. **a** The angular momentum $\mathbf{J}$ precess around the electric field axis z, while at the same time the permanent electric dipole moment $\boldsymbol{\mu}_0$, which is aligned along the symmetry axis of the cluster, nutates around the angular momentum vector. **b** Due to this pression-nutation motion the permanent dipole moment is averaged along the $\mathbf{J}$-axis which then is projected onto E_z. $K\hbar$ and $M\hbar$ are the components of $\mathbf{J}$ along the cluster symmetry axis and E_z, respectively, and β and γ are the enclosed angels of $\mathbf{J}$ with the corresponding axes

cle and $|\mathbf{J}| = J\hbar$.[14] Furthermore, $\mathbf{J}$ is precessing around z, resulting in a effective permanent dipole component of

$$\mu_{z,i} = \mu_0 \cos\beta \cos\gamma = \mu_0 \frac{K \cdot M}{J^2} \tag{3.18}$$

along the laboratory z-axis for the ith particle (called μ_z from here on), with $M\hbar$ being the projected component of $\mathbf{J}$ along the electric field axis (see Fig. 3.3). The FOPT interaction energy is obtained by increasing the electric field strength to a finite value and including the dipole polarizability contribution (see Eq. 3.2), giving

$$\epsilon_i^{(1)} = -\mu_0 \frac{K \cdot M}{J^2} E_z - \frac{\alpha}{2} E_z^2. \tag{3.19}$$

With Eq. 3.19 the relation Eq. 3.5 can be used to find the corresponding deflection d_i of the ith particle. This calculation could be repeated for all clusters in order to calculate the FOPT ψ_1-profile from Eq. 3.6. This, however, is of little use in order to assist the understanding of the experimental observations. Therefore, we rather calculate the ensemble averaged influence of the applied electric field on the field free beam profile ψ_0. Within the weak field approximation this can easily be done, since we expect only little change in ψ_0 when turning on the electric field. As mentioned earlier (see Sect. 2.4) the field free profile ψ_0 can be described by a Gaussian-type intensity distribution. This distribution only shows a first and a second central moment and, hence, the change of these moments between $\psi_1(E_z)$ and $\psi_0(E_z = 0)$ is investigated first. In order to calculate the ensemble average we need to know the partition function of the clusters in the electric field. In general, this is a difficult task, when not considering a thermalized ensemble in the electric field. However, in the considered case, the clusters in the molecular beam can not equilibrate with the surrounding due to the absence of collisions with the buffer gas, other clusters or the surrounding walls. Therefore, upon creation in the cluster source with T_{nozzle}, a canonical ensemble is formed. The clusters in the ensemble are further cooled by the expansion nozzle and the subsequent supersonic expansion. Due to the fact that we study a rigid cluster, T_{rot} is temperature of interest for calculating the partition function. After forming the molecular beam we will assume that this partition function will not change during the experiment. This is called the adiabatic ensemble assumption and will be qualitatively validated and discussed in some more detail in Sect. 3.4.

Since we know $\epsilon_i^{(1)}$, the corresponding Stark-effect (see Eq. 3.3) and the partition function we can determine the mean beam deflection

---

[14] Even though we are introducing $\hbar$, we are still using the classical approximation, since J, K, and M are not restricted to be integers. We only have the restrictions, $J \geq 0$, $-J \leq M \leq J$ and $-J \leq K \leq J$.

$$d = \langle d \rangle = \frac{\int_0^\infty \int_{-J}^{J} \int_{-J}^{J} d_i \exp[-BJ^2/(k_b T_{rot})] dK dM dJ}{\int_0^\infty \int_{-J}^{J} \int_{-J}^{J} \exp[-BJ^2/(k_b T_{rot})] dK dM dJ}$$
$$= \frac{\sigma}{mv^2} \frac{\int_0^\infty \int_{-J}^{J} \int_{-J}^{J} (\mu_0 \frac{KM}{J^2} + \alpha E_z) \exp[-BJ^2/(k_b T_{rot})] dK dM dJ}{\int_0^\infty 4J^2 \exp[-BJ^2/(k_b T_{rot})] dJ} \tag{3.20}$$

by simply calculating the classical average of the observable d_i. For this purpose the rotational constant B is introduced.[15] The first summand in Eq. (3.20) contains the antisymmetric functions K and M whose contributions will cancel when integrating in the interval $[J; -J]$. The polarizability term does not depend on J, K or M resulting in the simple relation

$$d = \frac{\sigma}{mv^2} \alpha E_z = \frac{\sigma}{mv^2} \langle \mu_z \rangle \tag{3.21}$$

connecting the average deflection (or average dipole moment in z-direction $\langle \mu_z \rangle$) of the molecular beam profile with the electric dipole polarizability α. The same calculation is easily performed for the second moment of d_i giving

$$\langle d^2 \rangle = \frac{\sigma^2}{(mv^2)^2} \frac{\int_0^\infty \int_{-J}^{J} \int_{-J}^{J} (\mu_0 \frac{KM}{J^2} + \alpha E_z)^2 \exp[-BJ^2/(k_b T_{rot})] dK dM dJ}{\int_0^\infty 4J^2 \exp[-BJ^2/(k_b T_{rot})] dJ} \tag{3.22}$$
$$= \frac{\sigma^2}{(mv^2)^2} \left(\frac{\mu_0^2}{9} + \alpha^2 E_z^2 \right)$$

and in a next step the second central moment

$$b^2 = \langle d^2 \rangle - \langle d \rangle^2 = \frac{\sigma^2}{(mv^2)^2} \frac{\mu_0^2}{9} = \frac{\sigma^2}{(mv^2)^2} \left(\langle \mu_z^2 \rangle - \langle \mu_z \rangle^2 \right) \tag{3.23}$$

is obtained with the help of Eq. (3.21).[16] The expression $\left(\langle \mu_z^2 \rangle - \langle \mu_z \rangle^2 \right)$ is equal to the variance of the dipole moment in z-direction. Consequently, the broadening of the molecular beam profile indicates the presence of a permanent electric dipole moment μ_0. Even if higher central moments exist, the beam deflections are that small, in the weak field limit, that no substantial alteration of the Gaussian-type shape of ψ_1 is observed. Therefore, we can conclude that the change of the molecular beam profile within the weak field approximation is dominated by an average beam deflection d (Eq. 3.21) and the beam broadening b^2 (Eq. 3.23). These can easily be used to extract the dielectric properties from experimental results. However, dielectric properties calculated from Eqs. 3.21 and 3.23 must be considered as a first estimate. For a more complete picture the effect of the cluster shape, T_{rot} and E_z have to be taken into account. Before doing this, it is instructive to derive the dipole distribution function

[15] Throughout the book the rotational constants have the dimension of an energy. A conversion into wavenumbers, commonly used for rotational constants, is possible by multiplying with $100/(h \cdot c)$.

[16] The same result is obtained when using quantum mechanical instead of classical FOPT calculations [74].

that gives rise to the beam deflection Eqs. 3.21 and 3.23. The deflection part of the distribution function is easily found, since we saw that only deflections proportional to αE_z are allowed and, hence, the overall beam profile will be shifted proportional to αE_z. For the broadening part of the distribution function originating from the permanent electric dipole moment, we follow the approach of Bertsch et al. who derived this for a magnetic moment in a weak magnetic field [75]. As mentioned earlier the dipole moment can point in any direction as long as μ_z is between $\pm\mu_0$. However, not all of these states are equally populated and we must calculate

$$\rho(\mu_z) = \frac{\int_0^\infty \int_{-J}^{J} \int_{-J}^{J} \delta(\mu - \mu_z)\exp[-BJ^2/(k_b T_{\rm rot})]dKdMdJ}{\int_0^\infty 4J^2\exp[-BJ^2/(k_b T_{\rm rot})]dJ} \tag{3.24}$$

where μ_z are the allowed dipole moment values from Eq. 3.18 and $\delta(\mu - \mu_z)$ is the δ-function, giving one when $\mu = \mu_z$ and otherwise zero. Substituting Eq. 3.18 for M into Eq. 3.24 and performing a variable transformation will give an easily integrable expression, resulting in the FOPT distribution function

$$\rho(\mu_z) = \frac{1}{2\mu_0} \ln\left(\frac{\mu_0}{|\mu_z|}\right) \tag{3.25}$$

originating from a permanent electric dipole moment.[17] This result is illustrated in Fig. 3.4a and compared to numerical calculations which will be described in Sects. 3.4 and 3.5. Both distribution functions are very similar, only near $\mu_z = 0$ the FOPT approximation exhibits a singularity, which is due to the used classical (continuum) approximation. Hence, at least in the weak field limit, the simple FOPT approach is able to describe the rotational motion of a cluster in a electric field.

However, this simple FOPT analysis in general fails to describe experimental results for clusters under cryogenic expansion conditions [77–80]. Especially, the average beam deflection d can irregularly overshoot the predictions of Eq. 3.21 by a factor of up to two and as a consequence the obtained polarizability values from the measured beam deflections will be (erroneously) increased. This is due to the permanent electric dipole moment of some clusters for which the FOPT approach breaks down. When we use the typical experimental values of $\alpha \sim 200\,\text{Å}^3$, $\mu_0 \sim 1\,\text{D}$, $E_z = 1 \cdot 10^7$ V/m (see Chap. 2) and $T_{\rm rot} = 5\,\text{K}$ [81] the reader easily can verify that the ratio between interaction energy and rotational energy for the polarizability is only 0.02 while for a cluster with a permanent dipole moment it can become ~0.5. For FOPT we have assumed that the interaction energy is small compared to the rotational energy, so that the rotational motion is not influenced by the electric field. For the described situation this is not longer the case and the electric field induced alignment of a cluster with $\mu_0 \neq 0$ must be taken into account. This can be done by a second order perturbation theory (SOPT) energy correction

[17] An analytic dipole moment distribution function for a symmetric rotor in the low field limit can be obtained, too (Sascha Schäfer, private communications). However, the influence of this shape correction is rather small.

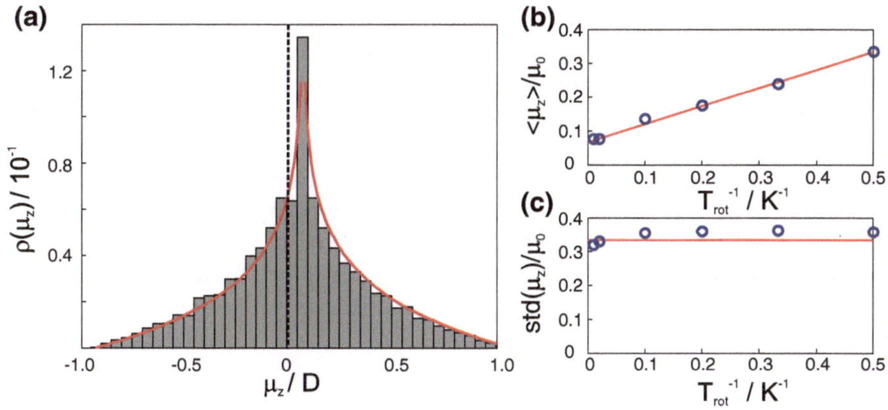

Fig. 3.4 Deflection properties of a rigid spherical rotor ($\mu_0 = 1$ D, $\alpha = 200$ Å^3 and $E_z = 1 \cdot 10^7$ V/m) in the weak field approximation derived from perturbation theory considerations. **a** Comparison between the analytic solution for the FOPT dipole moment distribution function (*red solid line*) and numerical calculations (*gray bars*, $T_{\mathrm{rot}} = 100$ K , see Sect. 3.5). The origin of the x-axis is highlighted by a *dashed black line*. **b** Normalized average dipole moment $\langle\mu_z\rangle/\mu_0$ as a function of T_{rot} according to SOPT (Eq. 3.27, *red solid line*) vis-a-vis with rotational dynamics simulations (*blue circles*, see Sect. 3.4). **c** FOPT predicts a nearly T_{rot}-independent behavior of the normalized standard deviation $\mathrm{std}(\mu_z)/\mu_0$ (*red solid line*) in good agreement with numerical simulations (*blue circles*)

$$\epsilon_i^{(2)} = \frac{\mu_0^2 E_z^2}{2B}\left[\frac{[J^2-K^2][J^2-M^2]}{J^3(2J+1)(2J-1)} - \frac{[(J+1)^2-K^2][(J+1)^2-M^2]}{(J+1)^3(2J+1)(2J+3)}\right] \tag{3.26}$$

for a symmetric rotor [82]. From this expression, obtained from quantum mechanical PT, the classical expression is found in the limit $J \gg 1$ [83] and gives

$$\alpha_{\mathrm{eff}} E_z = \left(\alpha + \zeta(\kappa)\frac{\mu_0^2}{k_{\mathrm{b}} T_{\mathrm{rot}}}\right) E_z = \langle\mu_z\rangle \tag{3.27}$$

in which the function $\zeta(\kappa)$ can take values between $(-1/3 + \pi/6)$ and $1/3$[18] and depends on $\kappa = I_a/I_c - 1$ (see Sect. 3.2 for the definition of the moments of inertia) [84, 85]. Hence, the experimentally measured effective polarizability α_{eff} consists of the electronic polarizability α and an additional term originating from the alignment of the cluster in the electric field. The latter is proportional to μ_0^2 and $1/T_{\mathrm{rot}}$ making this effect especially important for low rotational temperatures and high permanent dipole moments. In Fig. 3.4b the T_{rot}-dependence of a spherical rotor predicted by Eq. 3.27 and for numerical calculations (see Sect. 3.4) are compared. This

[18] For a spherical cluster the value of $\zeta(\kappa)$ is 2/9 and not 1/3 what would be expected in a canonical ensemble in thermal equilibrium.

clearly demonstrates the importance to incoorperate this T_{rot}-dependent effect for an analysis of the experimental beam deflections d. A similar SOPT analysis of the beam broadening is not possible since some of the integrals diverge [83]. However, if we compare the predictions from FOPT with numerical simulations (see Fig. 3.4c) it is obvious that for the extracted dipole moments and consequently for the beam broadening such a second order effect is only of minor importance.

Another effect that can be taken into account to correct the dipole moment value extracted from the experimental beam broadening, is the influence of the cluster structure. This means, the FOPT analysis can be performed for a symmetrical rotor in a weak electric field. The calculations are somewhat cumbersome and not really instructive. The main result is given by

$$b^2 = \left(\frac{\sigma}{mv^2}\right)^2 \frac{\mu_0^2}{3}\left[1 - 3\zeta(\kappa)\right] \tag{3.28}$$

where $\zeta(\kappa)$ is the same function as found for Eq. (3.27) [83, 86]. The effect of this correction changes the spherical FOPT result for typical cluster moments of inertia ratios by 5–30 % and can safely be neglected for most situations. A more detailed discussion of the influence of T_{rot} and the cluster shape on and the extracted dipole momemts from perturbation theory considerations can be found in Refs. [83, 86].

Here, we have shown that characteristic effects of electric beam deflection results can be understood by performing a perturbation theoretical analysis of a rigid almost spherical cluster in a weak electric field. Within this model a beam deflection d is connected to the polarizability α (Eq. 3.21) or to α_{eff} if a permanent dipole moment exists and T_{rot} is small. A beam broadening can be assigned to the presence of a permanent electric dipole moment μ_0. Furthermore, we have discussed the analytic dipole distribution function $\rho(\mu_z)$ (Eq. 3.24), which can serve as a guide for other more accurate methods. These methods will be described in the following sections.

3.4 Classical Rigid Rotor in an External Field

In Sect. 3.1 we have derived the force and deflection (Eqs. 3.3 and 3.5) in order to describe the experimental findings (Chap. 2). By doing that, we learned that beam deflection experiments at cryogenic temperatures probe the Stark effect of a rigid cluster with energy ϵ_i (Eq. 3.2) and that approximate solutions of the problem are found by applying perturbation theory (PT) methods (Sect. 3.3). From these considerations we were able to deduce the main characteristics of beam deflection experiments. Now, we want to find an (numerically) exact solution to the problem.

Therefore, the rotational motion of the rigid cluster in the electric field needs to be described. The most general way is to use quantum mechanics (Sect. 3.5) but here we want to treat the problem classically first. That the use of classical mechanics is justified for beam deflection experiments becomes clear by taking typical experimental conditions into account. Most beam deflection experiments are

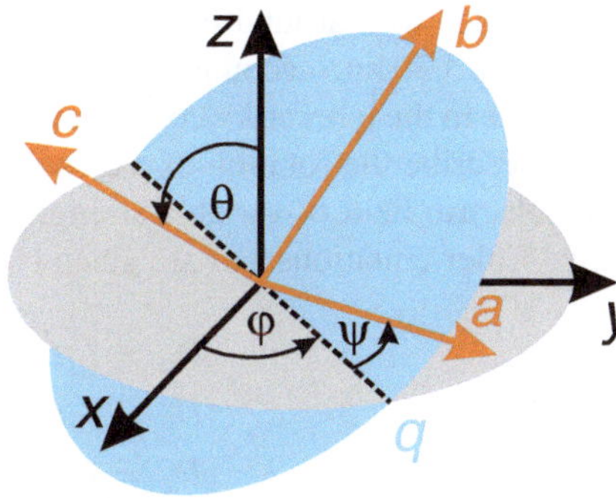

Fig. 3.5 Definition of the Euler-angels $\{\varphi, \theta, \psi\}$ used in this manuscript. The b Euler-angels φ, θ and ψ connect the space-fixed $\{x, y, z\}$ and the body-fixed $\{a, b, c\}$ coordinate systems by the successive rotations around z, q and c

performed in supersonic jets (Sect. 2.3). For these beams, it is well documented that the energy of the particle is not longer distributed equally over the different degrees of freedom [87, 88]. Hence, in supersonic beams T_{trans} (translation temperature), T_{vib} and T_{rot} are introduced to describe the amount of energy stored in the corresponding degree of freedom. In particular, the rotation is cooled effectively and T_{rot} is of the order of typically 5 K [81]. We assume that the mean moment of inertia value is of the order of $I = (10^{-43}–10^{-44})$ kg $\cdot$ m^2 and a semi-classical estimate

$$J \approx \sqrt{\frac{3k_{\text{B}}TI}{\hbar^2}} \approx (13 - 40) \tag{3.29}$$

shows that states up to high rotational quantum numbers $J$are occupied. Taking the correspondence principle into account, which states that classic mechanics is reproduced in the limit of large quantum numbers, the classical description of the rotational motion seems adequate. Only for very low temperatures or compounds with small I this approximation breaks down (see Sect. 4.1).

The first problem we encounter is that we observe and measure an object rotating in the laboratory frame (see Fig. 3.5) but we have defined the molecular properties, which describe the rotational motion and the interaction with the electric field, in the body-fixed coordinate system (Sect. 3.2). How can we connect the two coordinate systems? This is done by the Euler-angels $\{\varphi, \theta, \psi\}$ as shown in Fig. 3.5. By rotations around z, the intermediate axis q and c the two coordinate systems are interconverted.[19] These successive rotations can be used to transform a vector from the space-fixed into the body-fixed coordinate system via

$$\hat{S} = \begin{pmatrix} \cos\varphi\cos\psi - \sin\varphi\cos\theta\sin\psi & -\cos\varphi - \sin\varphi\cos\theta\cos\psi & \sin\varphi\sin\theta \\ \sin\varphi\cos\psi + \cos\varphi\cos\theta\sin\psi & \cos\varphi\cos\theta\cos\psi - \sin\varphi\sin\psi & -\cos\varphi\sin\theta \\ \sin\theta\sin\psi & \sin\theta\cos\psi & \cos\theta \end{pmatrix} \tag{3.30}$$

[19] There are different definitions of the Euler-angels. Here we follow the definitions used in [89].

or the other way around by using the inverse of $\hat{S}$. By taking the transformation matrix Eq. 3.30 into account, we can describe the rotational motion and all molecular quantities in the laboratory coordinate system. For the most general solution we must try to describe the rotation of a cluster of arbitrary shape, i.e. an asymmetric rotor, in an electric field of variable strength.[20] The equations describing this motion are called Euler-equations and are given by

$$\begin{aligned} D_a &= I_a \dot{\Omega}_a + (I_c - I_b)\Omega_b \Omega_c \\ D_b &= I_b \dot{\Omega}_b + (I_a - I_c)\Omega_a \Omega_c \\ D_c &= I_c \dot{\Omega}_c + (I_b - I_a)\Omega_a \Omega_b \end{aligned} \tag{3.31}$$

in which (D_a, D_b, D_c) and $(\Omega_a, \Omega_b, \Omega_c)$ are the components of the torque and angular velocity in the body-fixed coordinate system, respectively [89, 90].[21] Furthermore, we define $\mathbf{D} = \boldsymbol{\mu}_0 \times \mathbf{E}$ (neglecting α at this and adding its influence later) and in this way introduce $\boldsymbol{\mu}_0$ and $\mathbf{E}$ into Eq. 3.31. So far these equations are given in the $\{a, b, c\}$ coordinate system but experimentally observed are quantities in the space-fixed frame. By using Eq. (3.30), the Eq. (3.31) can be transformed but due to the rotational motion the angular momenta will change with time when observed from the laboratory coordinate system. Therefore, the three equations

$$\begin{aligned} \dot{\varphi} &= \Omega_a \frac{\sin\psi}{\sin\theta} + \Omega_b \frac{\cos\psi}{\sin\theta} \\ \dot{\theta} &= \Omega_a \cos\psi - \Omega_b \sin\psi \\ \dot{\psi} &= \Omega_c - \Omega_a \frac{\sin\psi\cos\theta}{\sin\theta} - \Omega_b \frac{\cos\psi\cos\theta}{\sin\theta} \end{aligned} \tag{3.32}$$

that connect the angular velocity in the moving frame with the time derivatives of the Euler-angles, are required. Thus, the task is to solve these six coupled differential equations numerically.[22] The situation is further complicated by the fact that these equation diverge for θ equal to 0 and π (see Eq. 3.32). This last mentioned problem is easily resolved by using quaternions [93]. Hence, this system of differential equations transformed into quaternions can be solved numerically by using a predictor-corrector algorithm or the Runge-Kutta method applying a static or varying electric field [83, 90, 94]. By choosing a set of initial values for $(\Omega_a, \Omega_b, \Omega_c)$ as well as $\{\varphi, \theta, \psi\}$ and using the parameters obtained from quantum chemical computations (Sect. 3.2) these calculations will yield a time-averaged electric dipole moment along the z-axis $\langle \mu_{z,i} \rangle_t$.[23] This result can be used to rewrite Eq. (3.5) to

---

[20] The same approach can easily be applied to spherical and symmetric rotors by making use of the symmetry of the moment of inertia tensor. Hence, this discussion can be considered as universally valid.

[21] The dot represents the time derivative of the corresponding quantity.

[22] In case of a symmetric rotor an analytical solution is found as described in [91, 92].

[23] The simulation time employed must be sufficient so that all quantities have converged.

$$d_i = \frac{\sigma}{m_i v_i^2} \left(\langle \mu_{z,i} \rangle_t + \alpha E_z \right) \tag{3.33}$$

which directly connects the deflection d_i with the calculated quantity $\langle \mu_{z,i} \rangle_t$ and the parameter α, obtained from quantum chemistry. As described before the experiment probes a whole ensemble of clusters. To account for this situation, a large but finite set of initial conditions is needed. In order to generate these states that adequately represent the experimental ensemble, we have to introduce an assumption that is crucial to all that follows. We will assume that the population of the different rotational levels is given by a canonical distribution formed in the cluster source with T_{rot} and that the distribution does not change when the clusters enter the electric field. Particularly, the second part of this statement needs further explanation. If the state of the cluster is assumed to be unchanged when entering the electric field the process is defined to be adiabatic. Hence, the change of the electric field must be slow compared to a typical time period of motion, in our case the rotational period τ [89]. An estimate for this period can be deduced from

$$\tau \approx \sqrt{\frac{4\pi^2 I}{3k_{\text{B}} T_{\text{rot}}}} \tag{3.34}$$

which gives 30–100 ps for the parameters used in Eq. (3.29). For a cluster traveling with $v = 10^3$ m/s and a field entrance region of 1 cm the penetration of the field will last 10 μs. Therefore, the field entrance is slow compared to the time scale of a typical rotation and the assumption of an adiabatic behavior of the cluster ensemble is fully justified. Consequently, the states required to perform the above outlined simulations are selected by an MC-Metropolis algorithm, creating a canonical ensemble for the chosen T_{rot} [18]. It is important to notice that no electric field is used for the MC search but this still yields the desired rotational level distribution as outlined above. Typically, this is done for 10^3–10^4 states. For each state the Eqs. 3.31 and 3.32 are solved to give $\langle \mu_{z,i} \rangle_t$ and the statistical weight is extracted from the MC calculations. This procedure yields the dipole moment distribution function $\rho(\langle \mu_{z,i} \rangle_t)$ which is called $\rho(\mu_{z,i})$ for clarity. By using Eqs. (3.33) and (3.6) the beam profile ψ_1 is obtained and the simulated profile can be compared to experimental findings.

For a discussion of typical results and the main characteristics of the obtained beam profiles T_{rot}-dependent simulations for a prolate, symmetric rotor with a permanent dipole moment of $\mu_c = 1$ D are shown in Fig. 3.6. An electric field of 10^7 V/m, a velocity of $v = 600$ m/s, a mass of 1200 amu, 4000 ensemble members and $I_a/I_b = 1$ as well as $I_b/I_c = 2$ have been used to perform the simulations. The position in the z-direction is called $p = z - z_0$ were z_0 is the maximum of the beam profile without field. From the discussion in Sect. 3.3 we know that α is responsible for a single sided deflection of the whole profile. The effect of α is easily incorporated by shifting the simulated profiles. However, in order to separate the effect of α and $\boldsymbol{\mu}_0$ we will discuss the hypothetial case $\alpha = 0$. In Fig. 3.6a the simulated profiles without (dashed line) and with electric field for a rotational

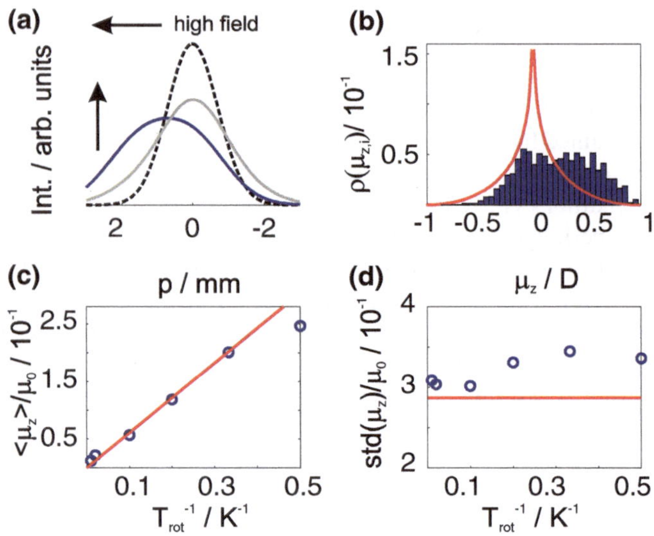

Fig. 3.6 Beam deflection profile simulation for a prolate, symmetric rotor with $\mu_c = 1$ D, $m = 1200$ amu, $v = 600$ m/s, $E_z = 10^7$ V/m and $I_a/I_b = 1$ as well as $I_b/I_c = 2$ for two T_{rot}. **a** Beam profile without (*dashed line*) and with electric field for 2 K (*solid line*) and 100 K (*gray solid line*). **b** Dipole moment distribution function for 2 K compared to predictions from PT for a spherical rotor. **c, d** Normalized mean value $\langle\mu_z\rangle$ (**c**, *circles*) and standard deviation std(μ_z) (**d**, *circles*) of the dipole moment as a function of T_{rot}^{-1} in comparison to predictions from PT (Sect. 3.3, *solid line*)

temperature of 2 K (solid line) and 100 K (gray solid line) are compared. For the high temperature simulation only a beam broadening is observed, while for the low temperature beam profile a deflection and broadening is recognizable. Furthermore, beside the broadening the shape of the 100 K-profile is very similar to ψ_0, which is described by a Gaussian function. In contrast to this observation is the form of the simulated low temperature profile. A clear deviation from the Gaussian profile shape is observed. The profile tails asymmetrically towards the direction of higher field strength. This can be rationalized if the corresponding dipole moment distribution functions are taken into account. Depicted in Fig. 3.6b is the low temperature profile (bars) for the symmetric rotor and the PT prediction (solid line) for a spherical rotor which is very similar to the numerical result for $\rho(\mu_{z,i})$ at 100 K (not shown). For 100 K, $\omega = \mu_0 E_z/(k_B T_{\mathrm{rot}}) \approx 0.02$ and, hence, much smaller than one. A similar calculation for 2 K gives $\omega \approx .1.21$. From this estimate it becomes apparent, that the dipole moment distribution function at 100 K is reproduced by predictions from PT but for 2 K the influence of the electric field on the rotational motion is not longer described accurately by PT but numerical results have to be taken into account. Shape-corrections of the PT distribution function as introduced in Sect. 3.3 are only responsible for a marginal variation of the herein shown results and do not change the main conclusion of the discussion. Furthermore, a close inspection of the dipole moment distribution functions shown in Fig. 3.6b, reveals the connection between $\rho(\mu_{z,i})$ and ψ_1. While the PT distribution function exclusively gives rise to a symmetric broadening, the 2 K simulation exhibits an asymmetric beam broadening and beam shift towards higher field strength. Hence, in order to further study the T_{rot}-dependence it is laborious to inspect more simulated beam profiles for various temperatures. It is more instructive to inspect the dipole moment distribution function and its corresponding characteristics as a function of T_{rot}. In particular, the mean value $\langle\mu_z\rangle$ and the standard deviation $\text{std}(\mu_z) = \sqrt{\langle\mu_z^2\rangle - \langle\mu_z\rangle^2}$ of $\rho(\mu_{z,i})$ give valuable information. These quantities can directly be compared to the PT-expressions describing the adiabatic polarization (Eq. 3.27) and magnitude of $\boldsymbol{\mu}_0$ (Eq. 3.23). The mean value and the standard deviation are depicted in Fig. 3.6c, d as a function of T_{rot}^{-1}. The numerical simulations for 2, 3, 5, 10, 50 and 100 K are shown as open circles while the PT predictions for a symmetric rotor (Eq. 3.27) are represented by solid lines. The smallest and largest x-data points correspond to the before discussed 2 and 100 K simulations, respectively. For the simulation at 100 K, $\langle\mu_z\rangle$ is close to zero, what is in good agreement with our previous qualitative discussion of the problem. With increasing inverse rotational temperature, the value of $\langle\mu_z\rangle$ increases linearly, too, consistent with predictions from SOPT given in Eq. 3.27 (Fig. 3.6c, solid line). Only at 2 K the predictions from PT and the simulation results differ. This is due to the large interaction energy between the dipole moment and electric field compared to the rotational temperature. Nevertheless, over a wide range of rotational temperatures the PT description of the adiabatic field polarization is sufficient to describe $\langle\mu_z\rangle$.

A somewhat different picture emerges from Fig. 3.6d for $\text{std}(\mu_z)$. As we have noted, the high temperature (100 K) results agree with predictions from PT (solid line) but for lower temperatures (around 5 K) the standard deviation from PT underestimates the broadening extracted from the numerical simulations. This is also in qualitative agreement with the observation we have made for the 2 K beam profile, for which $\rho(\mu_{z,i})$ significantly differs from the $\ln(\mu_0/\mu_{z,i})$ behavior obtained from FOPT (Fig. 3.6b). Therefore, we conclude that the rotational motion of a symmetric rotor in an electric field at high rotational temperatures can be reproduced by PT. For the mean dipole moment $\langle\mu_z\rangle$ the SOPT theory prediction agrees over a wide range of rotational temperatures with the results of the classical deflection simulations. However, at low rotational temperatures $\text{std}(\mu_z)$ clearly differs between PT and numerical classical calculations. Therefore, for these conditions the above outlined methodology must be used, in order to interpret the experimental results.

In a next step, we do not only want to assess the dependence of the molecular beam profile on ω but also on the shape of the cluster. In particular, the beam deflection profiles of asymmetric rotors are studied. For this purpose three model calculations are presented in Fig. 3.7. All simulation parameters are similar to those discussed previously for the symmetric rotor calculations with two exceptions. The ratio of the principle moments of inertia have been changed in order to study an asymmetric top. Here $I_a/I_b = 3/2$ and $I_b/I_c = 2$ were arbitrarily chosen. Additionally, the orientation of the dipole moment of 1 D is changed from $\mu_c = 1$ D (Fig. 3.7a, b) through $\mu_b = 1$ D (Fig. 3.7c, d) and finally to $\mu_a = 1$ D (Fig. 3.7e, f). The simulated beam profile of these asymmetric rotor calculations are shown in Fig. 3.7 for 2 K (solid line) and 100 K (gray solid line). In these model calculations all simulated 2 K beam profiles marginally depend on the dipole moment orientation and are qualitatively very similar. All show an asymmetric broadening which tails towards high field strength and a single sided deflection comparable with the 2 K symmetric rotor results (Fig. 3.6). Certainly, there is a quantitative difference between the 2 K beam profiles in Fig. 3.7a, c, e highlighting the sensitivity of the beam deflection method on the orientation of the permanent dipole moment. Nevertheless, qualitatively the simulated beam profiles show similar basic features. This is in vast contrast to the 100 K beam profile simulations. The three profiles are dramatically different. While the simulation for $\mu_c = 1$ D[24] shows a nearly symmetric broadening with only soft tailing and marginal beam deflection, similar to the case of a symmetric rotor (Fig. 3.7, a), the beam broadening for $\mu_a = 1$ D is clearly reduced and no single sided deflection is recognizable (Fig. 3.7e). Even more drastic is the effect when the dipole moment is aligned along the body-fixed b axis (Fig. 3.7c). In this case the beam broadening nearly disappears completely. Consequently, the orientation of the dipole moment only marginally influence the shape of the low temperature beam profiles but massively change the appearance of the 100 K results. In order to inspect this phenomenon in more detail the corresponding 100 K dipole moment distribution functions are depicted in Fig. 3.7b, d, f. For $\mu_c = 1$ D (Fig. 3.7b) $\rho(\mu_{z,i})$ exhibits a shape comparable to the PT distribution function for a spherical rotor

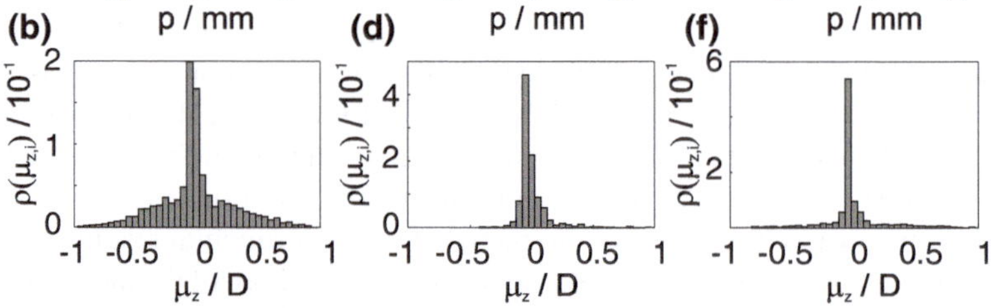

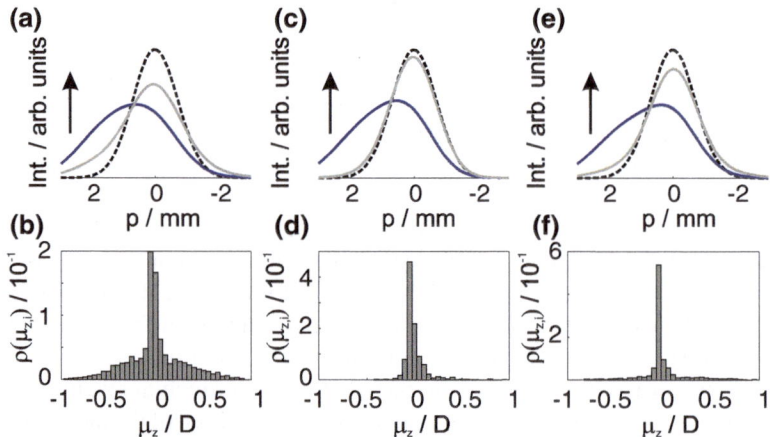

Fig. 3.7 **a, c, e** Beam profiles and **b, d, f** dipole moment distribution functions for an asymmetric rotor with $I_a/I_b = 3/2$ and $I_b/I_c = 2$ as well as (**a, b**) $\mu_c = 1$ D, (**c, d**) $\mu_b = 1$ D and (**e, f**) $\mu_a = 1$ D. **a, c, e** Beam profiles without (*dashed line*) and with electric field for 2 K (*solid line*) and 100 K (*gray solid line*). **b, d, f** $\rho(\mu_{z,i})$ for 100 K. The rest of the parameters are identical to those used for Fig. 3.6

[24] For a symmetric rotor this would be the symmetry axis.

(Sect. 3.3 and Fig. 3.6b). This is not quite surprising as the only difference to the above discussed symmetric rotor case are the principle moments of inertia. Above, we have shown that in the limit of $\omega \rightarrow 0$ the rotational motion of a symmetric rotor can be reproduced by PT. Unfortunately, this is not longer the case for the two other dipole moment orientations. As expected from the reduced broadening of the beam profiles, the dipole moment distribution functions clearly have reduced in width (Fig. 3.6d, f). An effect that is more dramatic for $\mu_b = 1$ D. What is the reason for this dipole moment orientation dependence, which additionally is influenced by ω? A simple picture that is able to explain this observation in the limit of $\omega \rightarrow 0$ emerges when the energy and angular momentum conservation

$$\epsilon_{\text{rot}} = \frac{J_a^2}{2I_a} + \frac{J_b^2}{2I_b} + \frac{J_c^2}{2I_c} \quad \text{and} \quad J^2 = J_a^2 + J_b^2 + J_c^2 \tag{3.35}$$

are taken into account [89]. In Eq. 3.35, $J = |\mathbf{J}|$ represents magnitude of the angular momentum vector and $\{J_a, J_b, J_c\}$ its vector components. In the $\{J_a, J_b, J_c\}$ coordinate system, the Eq. 3.35 define an ellipsoid and a sphere, respectively. For both equations to be simultaneously fulfilled, only the intersections of the ellipsoid and the sphere are allowed solutions to the problem.

These intersections produce closed loops as schematically depicted in Fig. 3.8 for the case of a symmetric prolate (a) and asymmetric (b) top. Since the permanent dipole moment is closely connected to the motion of the angular momentum vector for a rigid cluster, Fig. 3.8 can be used to qualitatively interpret the influence of the rotational motion on $\langle \mu_{z,i} \rangle_t$. For the symmetric rotor case (Fig. 3.8a) only orbits around the symmetry-axis c are allowed. Hence, a rotation according to this orbits will result in time-averaged quantities. Since there are no stable loops for the a- and b-axis, all properties defined along these directions will be zero. Only properties along the c-axis will be finite. This was expected for an symmetric top with $\mu_0 = \mu_c$. The situation changes significantly when examining the asymmetric rotor (Fig. 3.8b). Again orbits around c but additionally loops around a exist. The number of stable

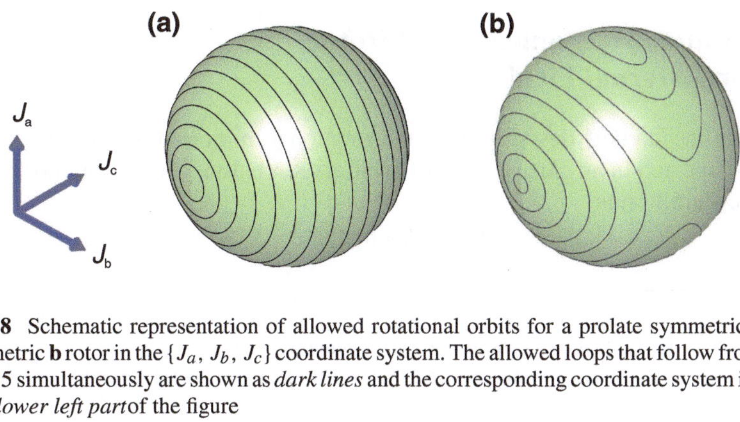

Fig. 3.8 Schematic representation of allowed rotational orbits for a prolate symmetric **a** and an asymmetric **b** rotor in the $\{J_a, J_b, J_c\}$ coordinate system. The allowed loops that follow from solving Eq. 3.35 simultaneously are shown as *dark lines* and the corresponding coordinate system is depicted in the *lower left part*of the figure

orbits around c and a depend on the ratio I_c/I_a and they determine how strongly a quantity defined along one of those axes averages over time. According to this qualitative discussion, a dipole moment along c as well as a does not average to zero and should be observable in experiment in the limit of $\omega \rightarrow 0$.[25] On the other hand, no orbits around the b-axis are allowed and, consequently, the dipole moment component μ_b along this direction will always average to zero. Therefore, μ_b can not be observed in beam deflection experiments in which $\omega \rightarrow 0$. Comparing these predictions to the presented 100 K results in Fig. 3.7 gives an intriguingly good agreement between the simple model and the numerical calculations. For the dipole moment aligned along c, a beam broadening is observed (Fig. 3.7a) whereas the beam profile for $\mu_a = 1$ D (Fig. 3.7e) is reduced in width. This clearly indicates the time-averaging effect of the rotational motion. Nevertheless, a reduced beam broadening is still observable. This is not longer the case when the dipole moment is oriented along the b-axis (Fig. 3.7c). Since no stable orbits around b exist, μ_b averages to zero and the resulting beam profile only shows a marginal drop in intensity. With this simple picture in hand we can qualitatively rationalize the beam deflection behavior of asymmetric rotors in the low field or high temperature limit. When E_z is increased or T_{rot} is lowered, as done in the 2 K simulations (Fig. 3.7a, c, e), this simple picture breaks down. The beam broadening is not longer determined by the rotational dynamics but an alignment of the rotor due to the increased interaction energy between the dipole moment and electric field takes place. Consequently, the 2 K beam profiles cannot be understood by only considering the magnitude of the dipole moment but its orientations in the body-fixed coordinate system and a numerically exact treatment of the rotational motion needs to be taken into account.[26]

The above discussed basic considerations highlight that beam deflection experiments not only depend on the magnitude of α and μ_0 but are sensitive to the dipole moment orientation, the cluster shape and the rotational dynamics. So far we only have considered the classical picture of a rigid rotor in an electric field. In a next step we want to treat the problem by means of quantum mechanics and compare the herein presented findings with the picture that emerges from the treatment in Sect. 3.5.

3.5 Quantum Mechanical Rigid Rotor in an External Field

A quantum mechanical description of the rotational motion of a rigid cluster with variable shape and in an electric field of arbitrary strength is, in general, possible by taking the Hamiltonian

$$\hat{H} = \hat{H}_{\text{rot}} + \hat{H}_{\text{Stark}} \tag{3.36}$$

---

[25] Situations exist in which the time-average of μ_a will be zero. However, this will be due to a special choice of I_c, I_a, J and ϵ_{rot} and is not true in general.

[26] These observations for an asymmetric rotor will be discussed in Sect. 4.2 for the case study of Ge-Clusters.

into account. The first term in Eq. 3.36 accounts for the rotational motion without electric field, while the electric field/cluster interaction is included in the $\hat{H}_{\text{Stark}}$-operator [82, 95]. To describe the rotational motion quantum mechanically, the angular momentum operators $\{\hat{J}_a, \hat{J}_b, \hat{J}_c\}$ need to be introduced into the classical expression for the rotational energy (see Eq. 3.35 and Ref. [89]), giving

$$\hat{H}_{\text{rot}} = \frac{\hat{J}_a^2}{2I_a} + \frac{\hat{J}_b^2}{2I_b} + \frac{\hat{J}_c^2}{2I_c} \tag{3.37}$$

for the $\hat{H}_{\text{rot}}$-operator. Similarly, the dipole moment in the classical Stark interaction term, of the form $-\boldsymbol{\mu} \cdot \mathbf{E}$, could be replaced by the corresponding dipole moment operator. The dipole moment, however, is defined in the molecular coordinate system. Hence, in order to calculate an experimentally measurable quantity, we need to transform the dipole moment operator into the laboratory coordinate system. In contrast to Eq. 3.37, where this can easily be achieved by using the basic relation $\hat{J}_a^2 + \hat{J}_b^2 + \hat{J}_c^2 = \hat{J}_x^2 + \hat{J}_y^2 + \hat{J}_z^2 = \hat{J}^2$ and the shift operators [82], the transformation operators $\hat{S}_{ij}$ (i and j can be $\{a, b, c\}$ and $\{x, y, z\}$, respectively) are required. These are the quantum mechanical analog of the matrix elements in Eq. 3.30. Taking this fact into account and constraining the discussion to cases in which the electric field is exclusively aligned along z, the expression

$$\hat{H}_{\text{Stark}} = -\mu_a E_z \hat{S}_{az} - \mu_b E_z \hat{S}_{bz} - \mu_c E_z \hat{S}_{cz} \tag{3.38}$$

is obtained. The task is to find the Eigenvalues and Eigenfunctions for the operator defined by the Eqs. 3.36–3.38. This, however, is not possible analytically, even for $E_z = 0$ [82, 95]. Therefore, we need to expand the unknown Eigenfunction in the basis of symmetric rotor Eigenfunctions $|J, K, M\rangle$. These functions are characterized by the quantum numbers J, K and M. The first can be connected to the length of the angular momentum vector. The latter two correspond to the projected angular momentum component along the c- and z-axis, respectively. In order to construct the Hamilton-matrix in this basis, the corresponding matrix elements are required. From the basic relations [82]

$$\begin{aligned}
\langle J, K, M|\, \hat{J}^2 \,|J, K, M\rangle &= J(J+1)\hbar^2 \\
\langle J, K, M|\, \hat{J}_c \,|J, K, M\rangle &= K\hbar \\
\langle J, K, M|\, \hat{J}_z \,|J, K, M\rangle &= M\hbar \\
\langle J, K \pm 1, M|\, (\hat{J}_a \mp i\hat{J}_b) \,|J, K, M\rangle &= \sqrt{J(J+1) - K(K \pm 1)}\hbar
\end{aligned} \tag{3.39}$$

the only non-zero matrix elements for the $\hat{H}_{\text{rot}}$-operator are given by

$$\begin{aligned}
\langle J, K, M | \hat{H}_{\text{rot}} | J, K, M \rangle &= \frac{\hbar^2}{4}\left(\frac{1}{I_a} + \frac{1}{I_b}\right)\left[J(J+1) - K^2\right] + \frac{\hbar^2}{2I_c}K^2 \\
\langle J, K \pm 2, M | \hat{H}_{\text{rot}} | J, K, M \rangle &= \frac{\hbar^2}{8}\left(\frac{1}{I_a} - \frac{1}{I_b}\right)[J(J+1) - (K \pm 1)(K \pm 2)]^{1/2} \\
&\quad [J(J+1) - K(K \pm 1)]^{1/2}.
\end{aligned} \tag{3.40}$$

More elaborate but still straightforward algebraic calculations will give the matrix elements of $\hat{H}_{\text{Stark}}$. All non-zero matrix elements, resulting from the interaction between the permanent dipole moment and the electric field along z, are given by

$$\begin{aligned}
\langle J, K, M | \hat{H}_{\text{Stark}} | J, K, M \rangle &= -\frac{K \cdot M}{J(J+1)}\mu_c E_z \\
\langle J+1, K, M | \hat{H}_{\text{Stark}} | J, K, M \rangle &= -\frac{\sqrt{(J+1)^2 - K^2}\sqrt{(J+1)^2 - M^2}}{(J+1)\sqrt{(2J+1)(2J+3)}}\mu_c E_z \\
\langle J, K \pm 1, M | \hat{H}_{\text{Stark}} | J, K, M \rangle &= -\frac{M\sqrt{J(J+1) - K(K \pm 1)}}{2J(J+1)}\mu_a E_z \\
&\quad \pm i\frac{M\sqrt{J(J+1) - K(K \pm 1)}}{2J(J+1)}\mu_b E_z
\end{aligned} \tag{3.41}$$

$$\begin{aligned}
\langle J+1, K \pm 1, M | \hat{H}_{\text{Stark}} | J, K, M \rangle &= -\frac{\sqrt{(J+1)^2 - M^2}\sqrt{(J \pm K + 1)(J \pm K + 2)}}{2(J+1)\sqrt{(2J+1)(2J+3)}}\mu_a E_z \\
&\quad -i\frac{\sqrt{(J+1)^2 - M^2}\sqrt{(J \pm K + 1)(J \pm K + 2)}}{2(J+1)\sqrt{(2J+1)(2J+3)}}\mu_b E_z
\end{aligned}$$

and the corresponding formulas where the plus sign has been replaced with a minus [82]. For an in-depth discussion of this topic and derivation of all matrix elements, the reader is referred to Refs. [82, 95].

By inspecting the Eqs. 3.40 and 3.41 the qualitative difference between a symmetric and asymmetric rotor becomes apparent. In the symmetric rotor case, the dipole moment is exclusively aligned along the symmetry axis c and $I_a = I_b$.[27] Without electric field only the first term in Eq. 3.40 is non-zero and $|J, K, M\rangle$ is an Eigenfunction of $\hat{H}$, as expected. Switching on the electric field will result in a shift in energy of the $|J, K, M\rangle$ states and a mixing of states with J and $J \pm 1$ (first two terms in Eq. 3.41). For an asymmetric rotor, their are two distinct cases that need to be considered. The first is when the dipole moment is still fixed along the c-axis but all principle moments of inertia are different. For this case, K and $K \pm 2$ states interact, only changing the absolute energy of the rotational levels but not the behavior in the electric field. However, if μ_a and μ_b are not zero, the rotational motion of the

[27] Here we only consider the case of a prolate symmetric rotor. For an oblate top similar relations can be obtained but the moment of inertia relation $I_c = I_b$ must be used.

cluster in an electric field will change (see last two terms in Eq. 3.41) compared to the symmetric rotor case.

After this qualitative discussion we want to find the numerically exact solution of $\hat{H}$ in the $|J, K, M\rangle$-basis, in order to compare the results for quantum mechanical symmetric and asymmetric rotors as well as to the prediction from classical mechanics (Sect. 3.4). The procedure is briefly described in the following. A detailed description can be found in Refs. [83, 96, 97]. For the most general case of an asymmetric rotor, only M remains as a "good" quantum number in the electric field, since the different J and K states can mix. Hence, for a fixed M value the $\hat{H}$-matrix is constructed using different J-states (a maximum value $J_{\max} = M + 20$ is used here) in the $|J, K, M\rangle$-basis and all $(2J + 1)$ possible K-states. The resulting matrix is diagonalized for different field strength E_z and therefrom Eigenvalues and Eigenfunctions are obtained as a function of E_z. For the symmetric rotor case, the problem is simplified due to the fact that only J and $J \pm 1$ states interact. In this case, the quantum numbers K and M remain valid. The $\hat{H}$-matrix will have a tridiagonal form, what allows to use a basis set including all J-states up to $100 \times k_b T_{\text{rot}}$. After diagonalizing the matrix, the Eigenvalues can be extracted. No matter if a symmetric or asymmetric rotor is considered, both procedures will give $\epsilon_i(E_z)$, i.e. the energy of the state $|i\rangle$ as a function of the electric field E_z. When changing the electric field to $E_z + \Delta E_z$, the energy will change to $\epsilon_i + \Delta\epsilon_i$. For sufficiently small ΔE_z (equivalent to $\Delta E_z/E_z \ll 1$), the relative energy change $\Delta\epsilon_i/\epsilon_i$ will be smaller than one[28] and the relation $\Delta\epsilon_i/\Delta E_z = -\mu_{z,i}$ will hold true. Hence, from calculating $\epsilon_i(E_z)$, the dipole moment $\mu_{z,i}$ of the state $|i\rangle$ in the direction of the z-axis is obtained. Combined with α, obtained from quantum chemical computations (see Sect. 3.2), and by using Eq. (3.5), the prediction of the corresponding d_i values is possible. In order to simulate the beam profile ψ_1 using Eq. (3.6), we need to calculate the distribution function $\rho(d_i)$ or equivalently $\rho(\mu_{z,i})$. This is done by using the energy of the state $|i\rangle$ without electric field $\epsilon_i^{(0)}$, in a Boltzmann-type distribution function, taking T_{rot} into account. This procedure will allow to fully describe the beam deflection behavior of a rigid cluster by means of quantum mechanics.

However, we want to start with a discussion of the field dependence of the energies ϵ_i, so called Stark-diagrams, from which the dipole moment distribution functions can be extracted. In particular, this will help to qualitatively understand the differences between symmetric and asymmetric rotors. First, we want to inspect the Stark-diagram of a symmetric rotor shown in Fig. 3.9. To make the discussion universally valid, the reduced quantities $\delta = \mu_0 E_z/B$ (where B is the rotational constant), ϵ/B and $\chi = 1/B \cdot d\epsilon/d\delta$ have been used. As a starting point, we can compare the numerically exact results for the arbitrarily chosen $|2, 2, -1\rangle$-state as a function of δ, shown in Fig. 3.9a+b, with predictions from PT (see Sect. 3.3). In the electric field this state mixes with $|1, 2, -1\rangle$ and $|3, 2, -1\rangle$. For the sake of simplicity we will continue to call this state $|2, 2, -1\rangle$. Some approximate energy expressions of the Stark-diagram of $|2, 2, -1\rangle$ are given by FOPT (Eq. 3.19), SOPT (Eq. 3.26), third order perturbation theory (TOPT, Eq. 3.42) or the high field pendulum approximation (Eq. 3.43) [98]. The TOPT and pendulum energy expressions are given by

[28] This is an assumption which breaks down for some asymmetric rotors as discussed below.

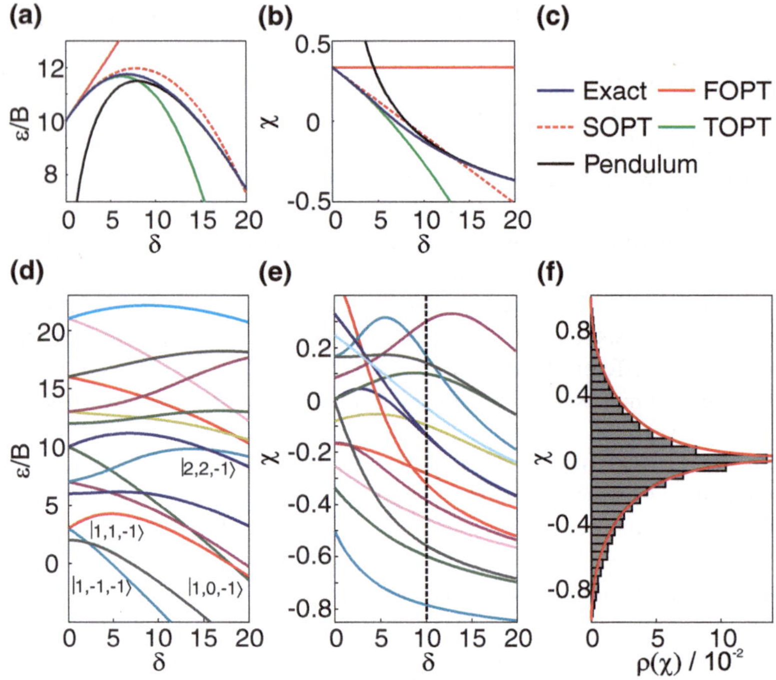

Fig. 3.9 **a** ϵ/B and **b** $\chi = 1/B \cdot d\epsilon/d\delta$ as a function of δ for the $|2, 2, -1\rangle$ state of a prolate symmetric rotor. The E_z-dependence of this state is compared to predictions from PT (color specification for a+b in **c**). Results of the numerical calculations for **d** ϵ/B and **e** χ for the 15 quantum states with $M = -1$ which are lowest in energy as a function of δ. Some $|J, K, M\rangle$-states are labeled. **f** Histogram of the symmetric rotor χ-distribution function (*gray*) for $\delta = 10$ and $k_B T_{rot} = 250 \cdot B$ in comparison to the FOPT distribution function (*red*). The rest of the parameters are equivalent to those for the classical symmetric rotor simulations (Sect. 3.4)

$$\epsilon_i^{(3)} = \frac{\mu_z^3 E_z^3}{4B^2}\left[\left(\frac{K \cdot M}{J(J+1)} - \frac{K \cdot M}{(J+1)(J+2)}\right)\left(\frac{\left[(J+1)^2 - K^2\right]\left[(J+1)^2 - M^2\right]}{(J+1)^4(2J+1)(2J+3)}\right)\right.$$
$$\left. + \left(\frac{K \cdot M}{J(J+1)} - \frac{K \cdot M}{J(J-1)}\right)\left(\frac{\left[J^2 - K^2\right]\left[J^2 - M^2\right]}{J^4(2J-1)(2J+1)}\right)\right] \tag{3.42}$$

and

$$\epsilon_i^{(\text{pendulum})} = \sqrt{2\mu_z E_z B}(2J - |K+M| + 1) - \mu_z E_z + \left(\frac{I_b}{I_c} - 1\right) B K^2$$
$$+ KMB + \frac{3(K-M)^2 - 3 - (2J - |K+M| + 1)^2}{8} B, \tag{3.43}$$

respectively [98]. While the FOPT approximation is able to describe the $\delta \rightarrow 0$ behavior of ϵ/B (Fig. 3.9a), it completely fails for χ and for ϵ/B at higher δ. By adding higher order PT terms to the FOPT approximation, the prediction from PT can gradually be improved. SOPT (Eq. 3.26) reproduces the ϵ/B and χ behavior over a wide range of field strength, with only introducing small deviations from the numerical result. The TOPT (Eq. 3.42) method is more accurate than SOPT up to $\delta \approx 6$ (the difference between SOPT and TOPT is small in this range) but then the curvature of the state is overestimated. The results from the Stark pendulum (Fig. 3.9a+b) asymmetrically approaches the numerical results for high δ but of course fails to reproduce the Stark-diagram at low field strength. Hence, the different regimes of the Stark-diagram of the $|2, 2, -1\rangle$-state can be approximated by employing different PT methods. In order to understand how the dipole moment distribution functions can be extracted from these diagrams, the reduced energy and χ of the 15 states with lowest energy and $M = -1$ are shown in Fig. 3.9d, e. First, we note from a close inspection of Fig. 3.9d, that crossings of the energy levels are allowed, since K and M states do not mix for a symmetric rotor. Hence, the energetic sequence of the rotational levels can change. In order to extract the dipole moment distribution function, $\chi = 1/B \cdot \mu_{z,i}/(\mu_0)$ needs to be considered, since this quantity is proportional to $\mu_{z,i}$. By extracting the different χ-values from Fig. 3.9e at $\delta = 10$ (dotted line) a small part of $\rho(\chi)$ is accessible, when the corresponding $\epsilon_i^{(0)}$-values (adiabatic field entrance) are taken into account. In Fig. 3.9f, the distribution function $\rho(\chi)$ (gray) from all states with up to an energy of $2500 \cdot B$ at $\delta = 10$ and for $k_B T_{rot} = 250 \cdot B$ is shown as a histogram. If we compare this distribution function with the predictions from spherical FOPT (Sect. 3.3), shown in the same figure as a red line, an intriguing agreement between the numerical and PT results becomes apparent. However, this is not surprising, since by choosing $k_B T_{rot}$ to be $250 \cdot B$ we set ω to $1/25$ and thus much smaller than one. Hence, we are discussing the high temperature or low field limit in Fig. 3.9f, for which FOPT is still valid. In this respect it is expected, that classical and quantum mechanical predictions of the distribution function are similar (see Sect. 3.4) in the $\omega \rightarrow 0$ limit. So far we have learned, how to infer all required information from Stark-diagrams and construct the distribution function $\rho(\chi)$. At least in the high temperature limit this quantum mechanical $\rho(\chi)$ is in agreement with predictions from PT (Sect. 3.3) and classical simulations (Sect. 3.4).

For low temperatures or high electric fields it has been shown that classical and quantum mechanical simulations give similar results within the numerical uncertainty but that the PT approach breaks down [83]. This again is expected, since ω becomes larger and PT is no longer an appropriate approximation.

What about the treatment of an asymmetric rotor by means of quantum mechanics? Can we rationalize the behavior of an asymmetric rotor in an electric field by inspecting the Stark-diagram? A Stark-diagram of an asymmetric rotor is shown in Fig. 3.10. By inspecting this picture, a dramatic difference to the symmetric rotor case (Fig. 3.9) becomes clear. Due to the mixing of all J and K states close in energy, no rotational states are allowed to cross. If two state come close in energy, they will repeal each other due to a strong mixing of the corresponding Eigenstates. In the case

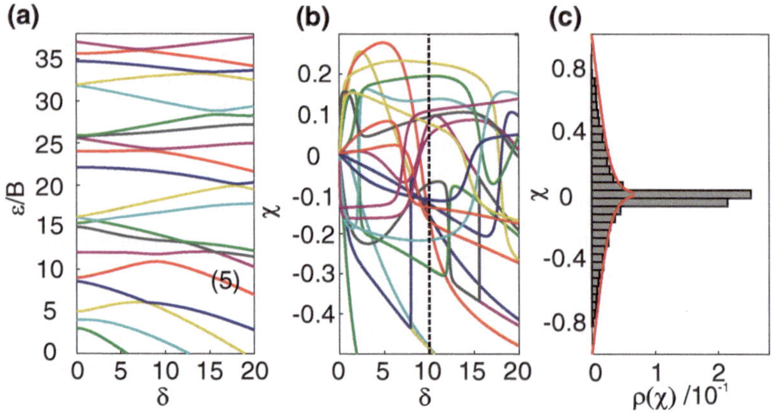

Fig. 3.10 **a** ϵ/B and **b** χ of the 20 lowest energy states of an asymmetric rotor with $M = -1$ as a function of $\delta = |\boldsymbol{\mu}_0|E_z/B$. **c** Dipole moment distribution function for the asymmtric rotor (*gray*) at $\delta = 10$ and $k_B T_{rot} = 250 \cdot B$ in comparison to predictions from FOPT (*red*). The 5th quantum state is highlighted by an attached label. The orientation of the dipole moment was fixed to $\mu_a = \sqrt{2/3}$ D and $\mu_c = \sqrt{1/3}$ D. All other parameters are identical to those used in Sect. 3.4 for an asymmetric rotor

of a symmetric rotor, where crossings are allowed, the crossing behavior is called diabatic. This is in contrast to the avoided or adiabatic crossings of the rotational levels of an asymmetric rotor, which characterize its Stark-diagram [47]. Even more dramatic is the effect of the avoided crossings on $\chi(\delta)$, as shown in Fig. 3.10b. Close to the avoided crossings, ϵ/B considerably change in a small δ-interval. This leads to irregular variations and discontinues of χ at these points and also influences $\rho(\chi)$, what is illustrated in Fig. 3.10c for $\delta = 10$ and $k_B T_{rot} = 250 \cdot B$ (gray histogram). The distribution function seems to consist of two components. One component is very similar to the symmetric rotor results, where the same temperature and electric field have been used in the simulations (Fig. 3.9f), and to the FOPT predictions (red). The second component seems to result from states with $\chi \approx 0$, i.e. states that have a small or zero dipole moment along the z-axis. Interestingly, this is very similar to the predictions of the classical high temperature simulations presented in Fig. 3.7. We can qualitatively rationalize this behavior by taking the effect of the avoided crossings into account. As shown in Fig. 3.10a, b some states undergo several avoided crossings. These states bounce back and forth and on average exhibit a small χ.[29] The effect of too many avoided crossings is a quenching of the Stark-effect of this state. This is in contrast to the diabatically crossing states of a symmetric rotor, which show no Stark-effect quenching. This statement is qualitatively confirmed by comparing the Figs. 3.9f and 3.10c. For this small selection of states contributing to $\rho(\chi)$, the density of states for most δ with $-0.1 \leq \chi \leq 0.1$ is higher in the asymmetric than in the symmetric rotor case. This is very similar to the observations of Xu et al.,

[29] At the point were the states cross this is not true, since the χ-values can vary considerably.

who discuss the behavior of clusters in magnetic fields [99]. Hence, the increased number of states with low χ originates from the avoided crossings, while the most states with large χ did not undergo many avoided crossings. For small T_{rot} or large ω mainly low J states are populated. These can exhibit large χ-values due to the small number of avoided crossings and, hence, a beam broadening is observable. This is in tune with the simulations presented in Sect. 3.4. By increasing T_{rot} more states with high J will be populated which are more likely to undergo avoided crossings, due to the increased density of rotational levels. Some of these states will show a quenched χ and, hence, the number of states with $\chi \approx 0$ will increase and the beam broadening will be reduced. This is in qualitative agreement with the classic beam profile simulations (Sect. 3.4) and serves as a simple model to explain the differences between the beam profiles of symmetric and asymmetric rotors.

Lastly, we will compare the qualitative model for the field free rotation of a classic rotor (Sect. 3.4) with numerical results for a quantum mechanical rotor. For this purpose the probability density of the z-axis in the molecular coordinate system for the symmetric rotor $|2, 2, -1\rangle$-state and the 5th-state of a asymmetric rotor (Fig. 3.10) are shown in Fig. 3.11 as a function of δ.[30] For the symmetric rotor at $\delta = 0$, the probability density is concentrated in a closed smeared out loop around the $-c$-axis. This is in accordance with the classic picture without electric field. On average, the c- and z-axis are opposing each other as expected for an $M = -1$ state. By increasing the electric field strength the closed loop around c is still present but the relative orientation is gradually inverted. Due to the applied electric field a brute force orientation [101] of the rotor takes place, until the c- and z-axis, on average, point in the same direction. This is in complete contrast to the arbitrarily chosen asymmetric rotor state. The shown 5th-state (Fig. 3.11) exhibits a probability density that is symmetric with respect to the a- and c-axis but no closed loops, neither around a or c, are present. This will result in a zero Stark-effect at $\delta = 0$ (see Fig. 3.10c). In our simplified classic picture (Sect. 3.4), however, an asymmetric rotor with μ_c and/or $\mu_a \neq 0$ shows a finite Stark-effect at very low field strength. This contradiction is

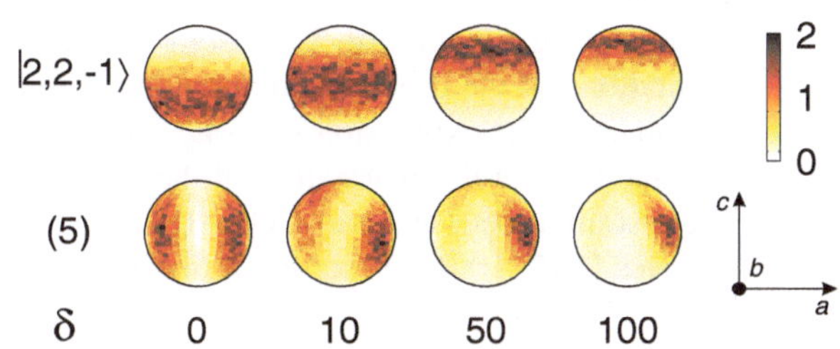

Fig. 3.11 Probability density of the z-axis in the molecular coordinate system for (*upper*) the symmetric $|2, 2, -1\rangle$-state (Fig. 3.9) and (*lower*) the 5th-state of an asymmetric rotor (Fig. 3.10) as a function of δ. The color code for of the probability density and the orientation of the molecular coordinate system are depicted on the *right side* of the figure

[30] See Ref. [100] for an in-depth discussion and definition of the $|J, K, M\rangle$-Eigenfunctions.

resolved for finite field strength (see for example Figs. 3.10c and 3.11). Under the influence of the electric field, the appearance of the probability density starts to change. From an opposed relative orientation of the c- and z-axis, giving rise to a positive Stark effect. The situation changes to an field induced orientation of the z-axis along the orientation of the dipole moment ($\mu_a = \sqrt{2/3}$ D, $\mu_c = \sqrt{1/3}$ D) of the asymmetric rotor.

Besides the behavior of an asymmetric rotor at very low field strength, the classic and quantum mechanical description of a rigid rotor in an electric field give similar results. Additionally, in the high temperature or low field limit these numerical procedures give results that are in tune with PT calculations. Which method actually can be used for a particular problem strongly depends on the experimental conditions and the system under study. While PT methods should only be used to interpret experiments with nearly effusive molecular beams, i.e. high rotational temperatures, or low electric fields, the time consuming numerical procedures are required for hard supersonic expansions, i.e. low rotational temperatures, and can be applied to any field strength. In the case of very low rotational temperatures or light elements, only the quantum chemical procedures presented above can provide a reliable description of the rotational motion [102]. However, the above presented picture can break down if either the vibrational motions have to be taken into account or if the behavior of the cluster in the electric field becomes chaotic [96, 97].

3.6 Floppy Clusters

All above considerations deal with rigid clusters. The situation, however, changes dramatically when vibrations are excited or isomerizations take place. A general description of these non-rigid clusters is very demanding and, hence, we want to start the discussion with the limiting case of a "fluxional" or "floppy" cluster [76]. In this simple model, the permanent electric dipole moment rotates rapidly (compared to the typical experimental time scale of several μs) in the molecular coordinate system (without applied electric field) due to vibrations and isomerization processes.[31] In an electric field, the orientations of the dipole moment along the electric field are energetically favored (compare to the expression $-\boldsymbol{\mu} \cdot \mathbf{E}$). These orientations have a larger statistical weight and an overall polarization of the polar clusters will take place. This situation is very similar to a Langevin-Debye (LD) behavior, a text book example in statistical physics [105, 106]. Therefore, a first idea is to use LD-type expression of the form

$$\alpha_{\text{eff}} = \alpha + \frac{\langle \mu_0^2 \rangle}{3k_{\text{B}}T_{\text{int}}} \tag{3.44}$$

---

[31] Additionally, the polarizability will change due to an thermal expansion of the cluster [103, 104]. However, as we will see this effect is small compared to the influence of μ_0 and only is important for non-polar clusters.

to describe the deflection behavior of "floppy" clusters. In Eq. (3.44) the effective polarizability α_{eff}, that can be probed in experiment, consists of the electronic polarizability α and a contribution $\langle\mu_0^2\rangle/(3k_B T_{\text{int}})$ originating from the partial alignment of the polar clusters in the $\omega \to 0$ limit. The internal energy T_{int} is a measure for the energy available for vibrations and isomerization processes and $\langle\mu_0^2\rangle$ is the ensemble average of the permanent electric dipole moment. Since it is difficult to characterize T_{int} in experiment, commonly $T_{\text{int}} = T_{\text{nozzle}}$ is assumed (T_{nozzle} is the temperature of the expansion nozzle) [107, 108]. Mostly, T_{nozzle} is in a next step set equivalent to T_{vib}. Therefore, the low field behavior of "floppy" clusters roughly scales with T_{vib} and must not be confused with the adiabatic polarization model of rigid clusters (Sect. 3.3), which scales with T_{rot}. Unfortunately, there is a significant flaw of this model. In contrast to the LD model, in which polar molecules in an electric field equilibrate with the surrounding, the polar clusters in a molecular beam, even if the assumption of a canonical ensemble prior to the field entrance is valid, enter the electric field adiabatically and are not able to redistribute the internal energy due to the lack of collisions. Hence, we have to deal with a cluster non-equilibrium ensemble, for which no simple analytic solution has been found so far. On dimensional grounds the low field behavior still should scale with $\mu_0 E_z/T_{\text{int}}$ but this is not as universally valid as the LD expression for gases of polar molecules [84]. Therefore, we have to emphasize again, that this model is only valid if the orientations of the dipole moment in the molecular coordinate system change rapidly compared to the typical experimental time scale ([10^{-5}–10^{-4}] s). What deflection behavior will be observed for "floppy" clusters in experiment? From Eq. (3.44) it is obvious that the polarizability, i.e. the shift of the molecular beam profile, will increase when the clusters posses a permanent electric dipole moment. On the other hand, we have assumed that the orientations of the dipole moment will readjust so that on average it will point in the same direction as E_z. Therefore, in high temperature experiments, in which the discussed approximation is valid, the molecular beam profiles only exhibit an increased beam deflection but no beam broadening. This has been demonstrated in various case studies (for some examples see Chap. 4) and it has been shown that a connection between the increased beam deflection and the presence of a permanent electric dipole moment exist (which additionally scales with T_{nozzle}) [6, 109–111].

How can we quantitatively understand the beam deflection behavior of clusters with excited vibrations and isomerization for all temperatures and field strength? The simple answer is: A generally applicable methodology for clusters and at all temperatures does not exist, yet! However, several case studies have demonstrated how to access the dynamic behavior of clusters in molecular beam deflection experiments to extract the time or ensemble averaged polarization in the electric field. By using numerical search routines (like MC) the populated isomers and their statistical significance can be identified. These can be used in conjunction with Eqs. (3.5) and (3.6) to simulate the molecular beam profile [112–114]. All these investigations deal with organic compounds. Hence, the use of force fields can give reliable results with only littlecomputational cost. For most cluster systems no reliable empirical potentials exist (see the discussion in Sect. 3.2) and, hence, similar calculations for clusters need to be performed employing quantum chemical methods. At present these calculations are computationally too time consuming, what complicates the interpretation of high temperature beam deflection experiments of clusters. In a first attempt that tries to overcome these limitations for clusters, an empirical potential that has been fitted to quantum chemical results is employed for the isomer identification, followed by quantum chemical calculations to deduce the permanent electric dipole moments [115]. Nevertheless, this method can only be applied to small clusters. Besides the lack of sufficiently effective numerical routines to interpret the high temperature beam profile of clusters, a qualitative understanding of the adiabatic field entrance of vibrational excited clusters (apart from the "floppy" cluster approximation) is still missing. This is the reason why high temperature beam profiles are very often only interpreted within the "floppy" cluster model. Hence, at the moment, the only way to directly extract the molecular properties α and μ_0 or structural information from these experiments is to perform low temperature beam deflection measurements.

References

1. Hohm U (2000) Vacuum 58:117
2. Maroulis G (ed) (2006) Atoms, molecules and clusters in electric fields. Imperial College Press, London
3. Bonin K, Kresin VV (1997) Electric-dipole polarizabilities of atoms, molecules and clusters. World Scientific Publishing company, Singapore
4. Scheffers H, Stark J (1934) Phys Z 35:625
5. Schäfer S, Mehring M, Schäfer R, Schwerdtfeger P (2007) Phys Rev A 76:052515
6. Moro R, Rabinovitch R, Xia C, Kresin VV (2006) Phys Rev Lett 97:123401
7. Kremens R, Bederson B, Jaduszliwer B, Stockdale J, Tino A (1984) J Chem Phys 81:1676
8. Tikhonov G, Wong K, Kasperovich V, Kresin VV (2002) Rev Sci Instrum 73:1204
9. Vollmer M, Selby K, Kresin V, Masui J, Kruger M, Knight WD (1988) Rev Sci Instrum 59:1965
10. Xu X, Yin S, Moro R, de Heer WA (2005) Phys Rev Lett 95:237209
11. Kittel C (2005) Einführung in die Festkörperphysik. Oldenburg Verlag, München
12. Kroto HW, Heath JR, O'Brien SC, Curl RF, Smalley RE (1985) Nature 318:162
13. Johnston RL (2003) Dalton Trans 22:4193–4207
14. Wales DJ (2003) Energy landscapes. Cambridge University Press, Cambridge
15. Ferrando R, Jellinek J, Johnston RL (2008) Chem Rev 108:845
16. Li Z, Scheraga HA (1987) Proc Natl Acad Sci U S A 84:6611
17. Wales DJ, Doye JPK (1997) J Phys Chem A 101:5111
18. Metropolis N, Rosenbluth AW, Rosenbluth MN, Teller AH, Teller E (1953) J Chem Phys 21:1087
19. Iwamatsu M, Okabe Y (2004) Chem Phys Lett 399:396
20. Zhan L, Chen JZY, Liu WK (2006) Phys Rev E 73:015701
21. Deaven DM, Ho KM (1995) Phys Rev Lett 75:288
22. Sierka M (2010) Prog Surf Sci 85:398
23. Heiles S, Johnston RL (2013) Int J Quantum Chem 113:2091

24. Wales DJ, Scheraga HA (1999) Science 285:1368
25. Cleri F, Rosato V (1993) Phys Rev B 48:22
26. Sutton AP, Chen J (1990) Phil Mag Lett 61:139
27. Ho KM, Shvartsburg AA, Pan B, Lu ZY, Wang CZ, Wacker JG, Fye JL, Jarrold MF (1998) Nature 392:582
28. Jiang D, Walter M (2011) Phys Rev B 84:193402
29. Kwapien K, Sierka M, Döbler J, Sauer J, Haertelt M, Fielicke A, Meijer G (2011) Angew Chem Int Ed 50:1716
30. Heiles S, Logsdail AJ, Schäfer R, Johnston RL (2012) Nanoscale 4:1109
31. Szabo A, Ostlund NS (1996) Modern quantum chemistry: introduction to advanced electronic structure theory. Dover Publication Inc, New York
32. Szalay PG, Müller T, Gidofalvi G, Lischka H, Shepard R (2012) Chem Rev 112:108
33. Lyakh DI, Musiał M, Lotrich VF, Bartlett RJ (2012) Chem Rev 112:182
34. Parr RG, Yang W (1989) Density-functional theory of atoms and molecules. Oxford Science Publications, Oxford
35. Geerlings P, De Proft F, Langenaeker W (2003) Chem Rev 103:1793
36. Ferrando R, Fortunelli A, Johnston RL (2008) Phys Chem Chem Phys 10:640
37. Heiles S, Hofmann K, Johnston RL, Schäfer R (2012) ChemPlusChem 77:532
38. Bast R, Ekstrom U, Gao B, Helgaker T, Ruud K, Thorvaldsen AJ (2011) Phys Chem Chem Phys 13:2627
39. Born M, Oppenheimer R (1927) Ann Phys 389:457
40. Jensen F (2007) Computational Chemistry. Wiley, Chichester
41. Møller C, Plesset MS (1934) Phys Rev 46:618
42. Dyall KG, Knut Fægri J (2007) Relativistic quantum chemistry. Oxford Univsersity Press, Oxford
43. Cohen HD, Roothaan CCJ (1965) J Chem Phys 43:S34
44. Dalgarno A (1962) Adv Phys 11:281
45. Mitroy J, Safronova MS, Clark CW (2010) J Phys B At Mol Opt Phys 43:202001
46. Helgaker T, Coriani S, Jørgensen P, Kristensen K, Olsen J, Ruud K (2012) Chem Rev 112:543
47. Landau LD, Lifschitz EM (2007) Lehrbuch der theoretischen physik: quantenmechanik. Verlag Harri Deutsch, Frankfurt am Main
48. Hohenberg P, Kohn W (1964) Phys Rev 136:B864
49. Kohn W, Sham LJ (1965) Phys Rev 140:A1133
50. Brack M (1993) Rev Mod Phys 65:677
51. Sousa SF, Fernandes PA, Ramos MJ (2007) J Phys Chem A 111:10439
52. Cohen AJ, Mori-Sanchez P, Yang W (2012) Chem Rev 112:289
53. Łach G, Jeziorski B, Szalewicz K (2004) Phys Rev Lett 92:233001
54. Gugan D, Michel G (1980) Mol Phys 39:783
55. Thakkar AJ, Lupinetti C (2005) Chem Phys Lett 402:270
56. Ekstrom CR, Schmiedmayer J, Chapman MS, Hammond TD, Pritchard DE (1995) Phys Rev A 51:3883
57. Kümmel S, Berkus T, Reinhard PG, Brack M (2000) Eur Phys J D 11:239
58. Bowlan J, Liang A, de Heer WA (2011) Phys Rev Lett 106:043401
59. Soos Z, Tsiper E, Pascal R Jr (2001) Chem Phys Lett 342:652
60. Bendkowsky V, Heinecke E, Hese A (2007) J Chem Phys 127:224306
61. Diercksen GH, Sadlej AJ (1988) Chem Phys Lett 153:93
62. Hebert AJ, Lovas FJ, Melendres CA, Hollowell CD, Story TL Jr, Street K Jr (1968) J Chem Phys 48:2824
63. Filsinger F, Wohlfart K, Schnell M, Grabow JU, Küpper J (2008) Phys Chem Chem Phys 10:666
64. Solov'yov IA, Solov'yov AV, Greiner W (2002) Phys Rev A 65:053203

65. Aguado A, Vega A, Balbás LC (2011) Phys Rev B 84:165450
66. Lei Y, Zhao L, Feng X, Zhang M, Luo Y (2010) J Mol Struc Theochem 948:11
67. Schäfer S, Heiles S, Becker JA, Schäfer R (2008) J Chem Phys 129:044304
68. Jackson K, Yang M, Jellinek J (2007) J Phys Chem C 111:17952
69. Colwell SM, Murray CW, Handy NC, Amos RD (1993) Chem Phys Lett 210:261
70. Neese F (2009) Coord Chem Rev 253:526
71. Jensen L, Autschbach J, Schatz GC (2005) J Chem Phys 122:224115
72. Merritt JM, Bondybey VE, Heaven MC (2009) Science 324:1548
73. Aguado A, Largo A, Vega A, Balbás LC (2012) Chem Phys 399:252
74. Schäfer S, Assadollahzadeh B, Mehring M, Schwerdtfeger P, Schäfer R (2008) J Phys Chem A 112:12312
75. Bertsch GF, Yabana K (1994) Phys Rev A 49:1930
76. de Heer WA, Kresin VV (2010) Handbook of nanophysics: clusters and fullerenes. Taylor and Francis, Boka Raton
77. Knickelbein MB (2001) J Chem Phys 115:5957
78. Schäfer R, Schlecht S, Woenckhaus J, Becker JA (1996) Phys Rev Lett 76:471
79. Knickelbein MB (2003) J Chem Phys 118:6230
80. Beyer MK, Knickelbein MB (2007) J Chem Phys 126:104301
81. Lenzer T, Bürsing R, Dittmer A, Panja SS, Wild DA, Oum K (2010) J Phys Chem A 114:6377
82. Kroto HW (2003) Molecular rotation spectra. Dover Publications Inc., New York
83. Heiles S, Schäfer S, Schäfer R (2011) J Chem Phys 135:034303
84. Schnell M, Herwig C, Becker JA (2003) Z Phys Chem 217:1003
85. Bulthuis J, Becker JA, Moro R, Kresin VV (2008) J Chem Phys 129:024101
86. Bulthuis J, Kresin VV (2012) J Chem Phys 136:014301
87. Haberland H (1995) Clusters of atoms and molecules I. Springer, Berlin
88. Bergmann L, Schaefer C (1992) Experimentalphysik 5: vielteilchen-systeme. Walter de Gruyter Verlag, New York
89. Landau LD, Lifschitz EM (2007) Lehrbuch der theoretischen physik: mechanik. Verlag Harri Deutsch, Frankfurt am Main
90. Dugourd P, Antoine R, El Rahim MA, Rayane D, Broyer M, Calvo F (2006) Chem Phys Lett 423:13
91. Bertsch G, Onishi N, Yabana K (1995) Z Phys D 34:213
92. Dugourd P, Compagnon I, Lepine F, Antoine R, Rayane D, Broyer M (2001) Chem Phys Lett 336:511
93. Evans DJ (1977) Mol Phys 34:317
94. Allen MP, Tildesley DJ (1987) Computer simulations of liquids. Oxford Science Publications, Weinheim
95. Townes CH, Schawlow AL (1975) Microwave spectroscopy. Dover Publications Inc., New York
96. Abd El Rahim M, Antoine R, Broyer M, Rayane D, Dugourd P (2005) J Phys Chem A 109:8507
97. Antoine R, El Rahim MA, Broyer M, Rayane D, Dugourd P (2006) J Phys Chem A 110:10006
98. Maergoiz AI, Troe J (1993) J Chem Phys 99:3218
99. Xu X, Yin S, Moro R, de Heer WA (2008) Phys Rev B 78:054430
100. Zare R (1988) Angular momentum: understanding spatial aspects in chemistry and physics. Wiley, New York
101. Friedrich B, Herschbach D (1991) Z Phys D 18:153
102. Moro R, Bulthuis J, Heinrich J, Kresin VV (2007) Phys Rev A 75:013415
103. Kümmel S, Akola J, Manninen M (2000) Phys Rev Lett 84:3827
104. Gamboa GU, Calaminici P, Geudtner G, Koster AM (2008) J Phys Chem A 112:11969
105. Langevin P (1905) J Phys Theor Appl 4:678
106. Hill TL (1986) An introduction to statistical thermodynamics. Dover Publications Inc, New York
107. Hopkins JB, Langridge-Smith PRR, Morse MD, Smalley RE (1983) J Chem Phys 78:1627
108. Collings BA, Amrein AH, Rayner DM, Hackett PA (1993) J Chem Phys 99:4174

109. Carrera A, Mobbili M, Marceca E (2009) J Phys Chem A 113:2711
110. Götz D, Heiles S, Schäfer R (2012) Eur Phys J D 66:293
111. Rayane D, Antoine R, Dugourd P, Benichou E, Allouche AR, Aubert-Frécon M, Broyer M (2000) Phys Rev Lett 84:1962
112. Antoine R, Compagnon I, Rayane D, Broyer M, Dugourd P, Breaux G, Hagemeister FC, Pippen D, Hudgins RR, Jarrold MF (2002) J Am Chem Soc 124:6737
113. Compagnon I, Antoine R, Rayane D, Broyer M, Dugourd P (2002) Phys Rev Lett 89:253001
114. Dugourd P, Antoine R, Breaux G, Broyer M, Jarrold MF (2005) J Am Chem Soc 127:4675
115. Kast SM, Schäfer S, Schäfer R (2012) J Chem Phys 136:134320

Chapter 4
Case Studies

After discussing the theoretical fundamentals of the methodology in Chap. 3, it is time to introduce the various analysis procedures by means of some examples. The incentive is to demonstrate how electric deflection techniques can be utilized to understand the dielectric properties of clusters as a function of their size and composition. For now, molecular clusters and complexes are discussed, with particular interest towards such aggregates that are linked via hydrogen bonding. Since such systems often exhibit weak vibrational modes, their internal dynamics will not freeze out, even in the seeded supersonic beam. Thus, as a general rule, only an effective polarizability can be observed which is significantly increased due to a dipolar contribution compared to the electronic polarizabililty. In contrast, metal-organic complexes exhibit a much more rigid atomic framework and can therefore be regarded within reasonable approximation as rigid. The same is also applicable for metal and semiconductor clusters, especially in such supersonic beams, where the carrier gas is pre-cooled. Hence, during the analysis of the electric beam deflection profiles of these aggregates one can compare the different perturbative approaches with the classical or the quantum-mechanical simulations of a rigid rotor quite well. Thereby, one can not only determine the electronic polarizabilities and permanent dipole moments of the cluster, but can also track in many cases the geometric structure as a function of the cluster size. For the element clusters of group 14, however, greater deviations from the characteristics of a rigid atomic framework are noticeable with increasing nozzle temperature and also atomic mass. The influence of chemical composition on the dielectric properties will be discussed in the last section, which introduces both alloyed metal clusters and core-shell-particles. By examining the dielectric properties one can, in particular, determine the growth pattern of such aggregates more precisely alongside with their electronic structure. All subsequent experiments were carried out with a "two-wire" field geometry.

S. Heiles and R. Schäfer, *Dielectric Properties of Isolated Clusters*, SpringerBriefs in Electrical and Magnetic Properties of Atoms, Molecules, and Clusters,
DOI: 10.1007/978-94-007-7866-5_4,

4.1 Molecular Clusters and Complexes

(a) The study of the *p*-aminobenzoic acid dimer denotes a particularly interesting example for the application of electric deflection experiments. The dimer is introduced by means of a matrix supported laser desorption into a pulsed molecular beam using helium as a carrier gas [1]. Detection is carried out via two-photon ionization at 266 nm in a position sensitive time-of-flight mass spectrometer. The two monomers are held together via a double hydrogen bond, yielding a centro-symmetric, planar complex in its electronic ground state. Therefore, one expects for the rigid dimer in the molecular beam only a one-sided deflection, which is proportional to the square of the electric field strength and depends on the electronic polarizability. Indeed, only a deflection of the molecular beam in direction of higher field strengths is observed in the experiment (see Fig. 4.1). However, the observed susceptibilities are significantly larger than the quantum-chemically predicted value for the electronic polarizability. This result can be explained by considering that the complex loses its centro-symmetric geometry through thermal excited vibrations, which will induce a dipole moment. Interestingly, this dipolar contribution to the susceptibility is temperature independent. This unusual temperature behavior can be explained by means of a simple model, in which the complex of N atoms is described by $3N-6$ (independent) classic, harmonic oscillators. To do this, the dipole moment is expanded as a function of the normal coordinates $\mathbf{Q}_i$

$$\boldsymbol{\mu} = \sum_i \left[\left(\frac{\partial \boldsymbol{\mu}}{\partial \mathbf{Q}_i} \right) \mathbf{Q}_i \right], \tag{4.1}$$

and further one has to take into account that in the experiment an average value $\langle \mu^2 \rangle$ over all possible displacements, which are accessible at the corresponding temperature, is observed (see Eq. 3.44). A temperature independent contribution $\langle \mu^2 \rangle$ to the susceptibility arises due to the fact that the mean square displacements $\langle Q_i^2 \rangle$ increases proportionally with the temperature T in the canonical ensemble. This interpretation is confirmed by Monte-Carlo simulations, where it is shown that the average value $\langle \mu^2 \rangle$ does not necessarily increase strictly linear with the temperature [1]. This is presumably due to mechanical anharmonicities, but overall an approximately temperature-independent effective polarizability is obtained. Thus, the combination of Monte-Carlo simulations and experimental results give rise to a virtually quantitative agreement and a qualitative confirmation of the simple oscillator model (Fig. 4.1).

(b) The by far most important hydrogen bridge forming substance is water. For this reason, electric deflection experiments on water clusters are particularly important in order to understand how dielectric properties of water aggregates change as a function of their size. In spite of this interest, corresponding experiments on water clusters as well as on the water molecule have only recently been performed. There are different reasons for this: The ionization potential of the water molecule is so large that investigation can only be performed by utilizing the relatively insensitive

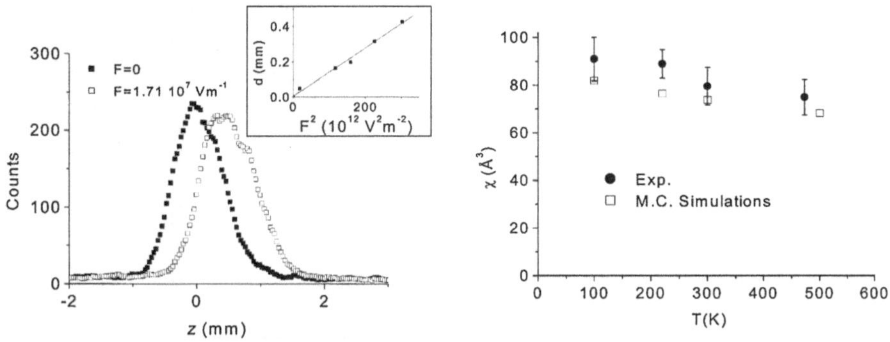

Fig. 4.1 The molecular beam profile of the *p*-aminobenzoic acid without (*filled squares*) and with (*open squares*) an applied electric field (here called F) is depicted on the *left hand side* of the figure [1]. The electrical field strength is $1.71 \cdot 10^7$ V/m. An one-sided deflection without broadening is evident. The deflection d is proportional to the effective polarizability with a quadratic field strength dependency as shown in the *inset*. On the *right hand side*, the experimentally observed susceptibility is illustrated as a function of the nozzle temperature (*filled circles*). In addition, it is shown that values of the electric susceptibility obtained via Monto-Carlo simulations only depend marginally on the internal temperature of the dimer. Reprinted figure with permission from Compagnon et al. [1]. Copyright 2002 by the American Physical Society

method of electron impact ionization. Further, with H_2O being an asymmetric rotor, a quantitative analysis of the deflection profiles is only possible by using non-perturbative procedures. Moreover, since H_2O exhibits the largest values of rotational constants of all stable molecules and thus even at room temperature only a few rotational states are occupied, a quantum mechanical description of the rotational dynamics within the electric field is necessary. Kresin et al. carried out a corresponding analysis assuming H_2O to be a rigid, asymmetric rotor [2]. They compared the resulting dipole moment distribution function with the measured deflection profiles. It was concluded that in the case of an asymmetric rotor, clusters which adiabatically enter the field experience an electric polarization, quite similar to what has previously been shown by second order perturbation theory for spherical and symmetric rotors (Eq. 3.27). The same research group investigated, for the first time, the electric deflection characteristics of $(H_2O)_N$ clusters ($n = 3$–18) (see Fig. 4.2) [3]. The water clusters were produced by a continuous gas aggregation source, whereupon supersonic cluster beams with and without helium as carrier gas were formed. Cluster detection was performed via electron impact ionization in a quadrupole mass spectrometer, where beam profiles were collected by using a movable slit. It is especially important to be sure that mass spectra are not contaminated by fragmentation processes. While the water molecule can still be regarded as a rigid rotor, this is not feasible in the case of water clusters, which exhibit very weak vibrational modes that are already thermally excited even in supersonic beam experiments. The structure of the water clusters in an electric field is therefore not rigid and the alignment of the electric dipole moment follows a random walk. The probability for the occurrence of a specific orientation is thus given in the statistical limit of the canonical

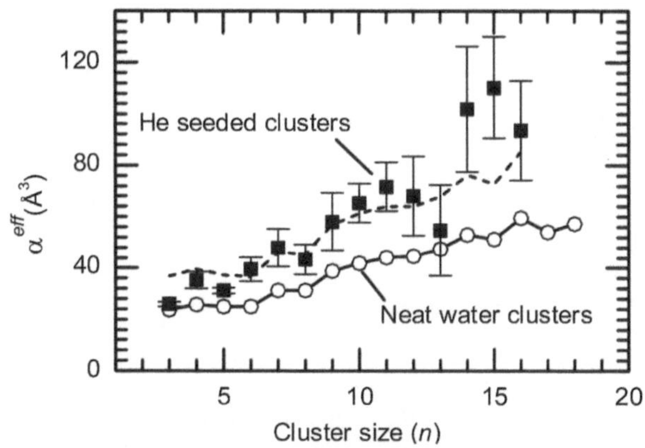

Fig. 4.2 The dependence of the effective polarizability α^{eff} as a function of cluster size N for $(H_2O)_N$-clusters and different expansion conditions [3]. Clusters formed with helium as a seeding gas show considerably larger values for α^{eff} in comparison to aggregates produced via expansion of the pure water vapor. This is due to a lower internal temperature of the seeded clusters. It is peculiar that in both cases the values of the effective polarizability increase almost linearly as a function of cluster size, that is each water molecule contributes, independent of cluster size, a very specific electronic and dipolar contribution to the electric susceptibility. Reprinted figure with permission from Moro et al. [3]. Copyright 2006 by the American Physical Society

distribution and influenced by the internal temperature of the clusters. Indeed, this behavior has been observed experimentally. The internal temperature of the clusters is a result of the expansion conditions and the subsequent cooling to 200 K by monomer vaporization. The observed effective polarizabilities are mostly identified by the additional dipolar contribution, where dipole moments averaged over all orientations are on the order of $(1.3 - 1.6)$ D. This behavior is qualitatively found in experiments with helium-seeded water cluster beams, too. Here, lower internal temperatures can be achieved and consequently the dipolar contribution to the susceptibility is further increased. A quantitative interpretation of the observed values for the averaged dipole moments by means of, for instance, Monte-Carlo simulations like in the case of the *p*-aminobenzoic acid dimer is still pending (Fig. 4.2).

(c) Marceca et al. carried out experiments with sodium doped $(H_2O)_N$-clusters [4]. The doped clusters were produced by means of a "pick-up" cell, where water cluster absorb sodium atoms while flying through a corresponding vapor atmosphere. Beam profiles were obtained by spatially scanning the molecular beam using an ionization laser. It is particularly interesting to analyze how the absorption of an additional atom influences the dielectric properties of small water clusters. It is known that during doping of small water clusters with sodium atoms an electron transfer occurs, and that the resulting sodium ion is surrounded in its first coordination shell by four water molecules and the electron is delocalized outside of this cavity. The addition of sodium gives therefore rise to a modified configuration of the water molecules within the cluster and induces an internal dipole. This behavior is also predicted quantum-chemically and has been observed for the first time in electric deflection experiments. The effective polarizabilities of the clusters doped with sodium are increased by about 100 Å^3 in comparison to pure water clusters. On the one hand, this is due to an increased electronic polarizability owing to the delocalized electron and on the other to an increased dipolar contribution, since the averaged dipole moments are now about $(2.5-3.4)$ D. Thus, despite of the restructuring influences of sodium, these observed clusters cannot be regarded as rigid under these experimental conditions.

(d) Electric deflection methods can also be used to investigate the dielectric properties of metal-organic complexes. For example, Broyer et al. performed deflection experiments to analyze transition metal complexes comprising of two benzene ligands [5, 6]. These complexes were produced via laser ablation of the transition metal in a helium atmosphere that contained a small quantity of benzene. This gas mixture was subsequently expanded to form a pulsed molecular beam containing the respective metal organic complexes. Two families of structures are observed here, depending on the transition metal: symmetric sandwich structures (D_{5h}) without a permanent electric dipole moment in the electronic ground state and polar, asymmetric complexes (C_{2v}) exhibiting dipole moments between $(1-2)$ D. Measured flight times obtained by using a position sensitive mass spectrometer with and without an electric field reveal instantly that while $Ti(C_6H_6)_2$ forms a symmetric sandwich compound, $Co(C_6H_6)_2$ is asymmetric and possesses a permanent electric dipole moment (see Fig. 4.3). The small and uniform shift of the flight times in the case of $Ti(C_6H_6)_2$ is caused by the electronic polarizability. The broadening of the flight times in the case of $Co(C_6H_6)_2$ indicates that the orientation of the dipole moment in the electric field is preserved at least in part. Therefore, the broadening of the flight times was simulated presuming that the complex can be regarded as a rigid rotor. Since $Co(C_6H_6)_2$ is a weak asymmetric rotor ($I_a \approx I_b$), its dipole moment is essentially only non-zero along the main axis a, and the broadening of the flight time distribution can be classically simulated employing an oblate, symmetric rotor model. The classical description is justified, since the experiment is carried out at room temperature and all rotational constants are smaller than 0.1 cm^{-1}. Taking quantum-chemically calculated rotational constants into account, a very good agreement with experimental data is obtained (solid curve). The dipole moment (0.7 ± 0.3) D extracted from the simulation is, however, significantly smaller than the value of 1.6 D obtained from density functional theory calculations (exchange-correlation functional B3LYP). Possible causes for this discrepancy are on the one hand an inadequate level of quantum chemistry and, more likely, on the other hand that the clusters are not entirely rigid, resulting in an observed broadening of the flight time distribution that is too small (Fig. 4.3).

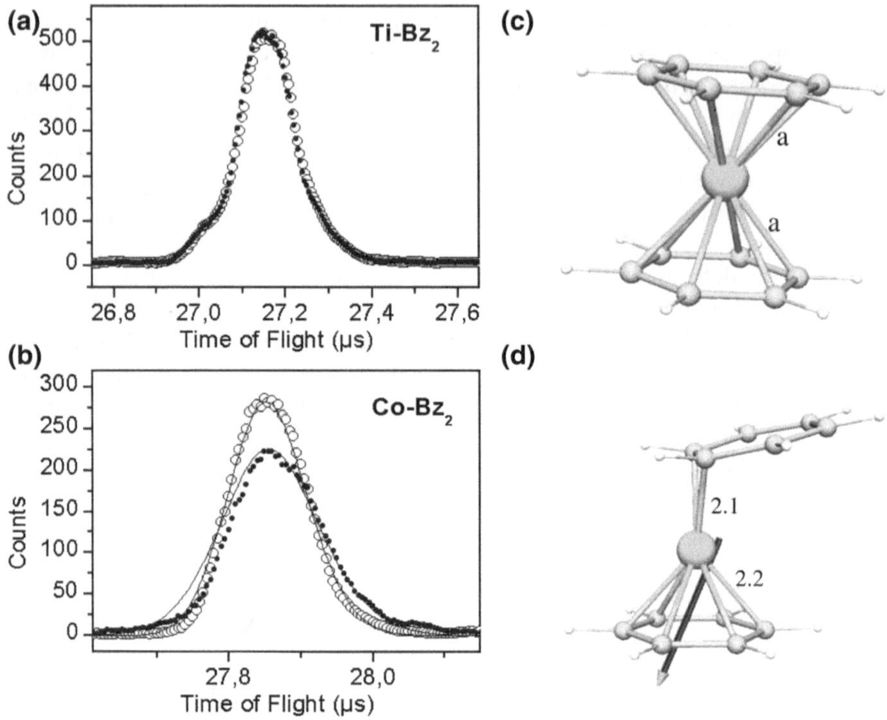

Fig. 4.3 The time-of-flight mass spectra of a symmetric and asymmetric metal-organic complex comprising of a transition metal and two benzene ligands, respectively **a** symmetric $Ti(C_6H_6)_2$, **b** asymmetric $Co(C_6H_6)_2$) [5, 6]. Structures predicted by quantum-chemical calculations (level of theory: B3LYP [7] with SDD-basis [8]), including some bonding distances in Å and the orientation of the electric dipole moment, are depicted in (**c**) for $Ti(C_6H_6)_2$ ($a = 2.29$ Å) and in **d** for $Co(C_6H_6)_2$ [5, 6]. In both time-of-flight spectra, *open circles* denote measurements without an electrical field and *filled circles* with an electric field strength of $1.51 \cdot 10^7$ V/m. In **a**, only a slight shift of the flight time is observed for $Ti(C_6H_6)_2$ consistent with a theoretically predicted symmetric structure. In **b**, a broadening of the profile is also found for $Co(C_6H_6)_2$ which is a result of the polar structure. The time of flight profile with an electric field is simulated (*continuous line*) in **b**, presuming a rigid, symmetric rotor ($A = B = 0.0190$ cm^{-1}, $C = 0.0502$ cm^{-1}) with a dipole moment of 0.7 D. Here, the limiting case of small electric fields and high rotational temperatures, respectively, ($\omega = \mu E/k_B T_{rot} \rightarrow 0$ and $1/T_{rot} \rightarrow 0$) is used. While the time-of-flight profile can be nicely reproduced, the theoretically predicted value of 1.6 D is more than twice as large as the value of (0.7 ± 0.3) D observed in experiment. Reprinted with permission from Broyer et al. [5]. Copyright 2007, Institute of Physics

4.2 Metal and Semiconductor Clusters (Group 14)

The application of electric deflection methods to determine the dielectric properties of metal and semiconductor clusters is showcased for the clusters of group 14. Within group 14 the strongest changes of the physical-chemical properties take place from the typical non-metal carbon to the heavy metal lead and thus the analysis of such clusters is specifically interesting to determine to what extent the properties of the macroscopic solids are reproduced in the small clusters.

(a) The deflection behavior of the fullerenes, that is the clusters of the lightest homologue of group 14, was studied for the first time by Broyer and coworkers [9]. A suitable target was vaporized by employing a Nd:YAG laser, the produced fullerenes were then expanded with helium as a carrier gas and were detected by photoionization in a position sensitive mass spectrometer. Only an electronic contribution to the polarizability is expected for the centro-symmetric C_{60} and C_{70} molecules and corresponding deflection profiles were also observed in the experiment. Polarizability values extracted from the beam shift are within the margin of error in good agreement with quantum-chemical predictions ((76.5 ± 8) Å^3 for C_{60} and (102 ± 12) Å^3 for C_{70}, respectively). Since, there are only weak intermolecular interactions between individual fullerene molecules in the solid state of C_{60} and C_{70}, respectively, and since their charge densities only change insignificantly in comparison to the free molecules, the Claussius-Mosotti-relation [10]

$$\frac{\alpha}{4\pi\epsilon_0} = \frac{3V}{4\pi}\frac{\epsilon(0)-1}{\epsilon(0)+2} \tag{4.2}$$

can be used together with the dielectric constants to determine the molecular polarizabilities. This is done by using the lattice constants of the fcc unit cells 14.17 Å (C_{60}) and 15.01 Å (C_{70}) to calculate the volume V of a free fullerene molecule. The values for the static dielectric constants $\epsilon(0)$ were obtained from optical experiments on thin fullerene films and are 3.61 (C_{60}) and 3.76 (C_{70}), respectively. Using these constants, values of 79 and 97 Å^3 are obtained for the polarizabilities. These agree within the uncertainty of measurement with the results of electric deflection experiments. Indeed, this illustrates that the intermolecular interactions between discrete fullerene molecules in the solid state are very weak, and on the other hand that the dielectric properties of the bulk can be predicted via observed polarizabilities of the isolated fullerenes.

(b) The chemistry of carbon is in view of the pronounced tendency to form double bonds drastically different from that of silicon compounds. Thus, one can expect that in the case of small silicon clusters a free surface with unsaturated valence electrons does not lead to multiple bonds but that instead the coordination sphere must reconstruct itself to minimize the surface (free) energy. Furthermore, small silicon clusters should neither exhibit fullerene-type structures, nor such structural motifs that are present in the bulk. Exactly this is experimentally observed (Fig. 4.4) and one is therefore inclined to presume that the dielectric properties of the silicon clusters are different from those of the macroscopic bulk. Units with trigonal prism symmetry represent a characteristic structural motif. For example, the structure of the Si_{18} isomer with the lowest energy is comprised of a trigonal prism and an antiprism that are each tricapped and interconnected with each other (inset, Fig. 4.4). Considering the elongated structure of this cluster it comes to no surprise that a huge dipole moment of more than 4 D is formed along the three-fold axis of rotation. This dipole moment becomes noticeable in the effective polarizability measured at

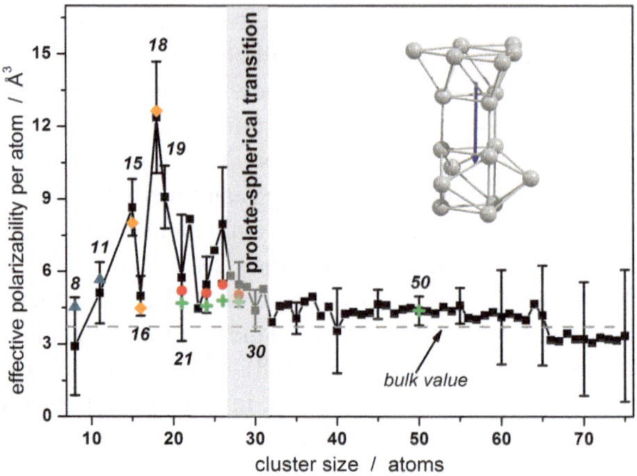

Fig. 4.4 Effective polarizabilities per atom of Si_N-Cluster at room temperature as a function of cluster size (*filled squares*) [11]. The silicon clusters were produced utilizing a pulsed laser vaporization source and were seeded with helium as a carrier gas. Beam profiles were obtained by means of a variable slit with and without an applied electric field. Effective polarizabilities are determined via the shift of beam profiles when the electric field is turned on employing the Langevin-Debye model (see Sect. 3.6 and Eq. 3.44). Larger error bars for clusters with $N > 60$ are a result of a smaller data set. The *dashed line* represents the polarizability of a small, dielectric sphere with the properties of α-Si (see Eq. 4.2). It is peculiar that some clusters exhibit values that are significantly larger than the mere electronic polarizability expected from the bulk. This can be attributed to the existence of polar cluster structures. As an example, the *inset* shows the theoretically predicted structure of Si_{18}, the *blue arrow* indicate the direction of the electric dipole moment. A markedly good agreement with experiment is obtained when theoretically calculated dipole moments are taken into account to calculate the additional dipolar contribution to the electronic polarizability. The theoretical predictions are shown for some cluster sizes by colored symbols (Ref. [12]: ▲ fig (a), Ref. [13]: ◆ fig (b), Ref. [14]: ● fig (c) [prolate structures] and ✚ fig (d) [compact structures]). Moreover, the transition from rather elongated to spherical structures is easily noticeable for silicon clusters with $N > 30$, since the polarizabilities per atom do not strongly fluctuate and are not increased significantly compared to the bulk value. Reprinted with permission from Götz et al. [12]. Copyright 2012, European Physical Society and Springer

room temperature (see Fig. 4.4). While an electronic polarizability of about 4 Å^3 per atom is predicted theoretically, a value of more than 12 Å^3 is observed [11]. Moreover, the existence of a permanent dipole moment for other cluster sizes is also evident from such increased effective polarizabilities. This behavior can be interpreted quantitatively by an additional dipolar contribution to the polarizability. The observed polarizabilities can therefore be explained by means of theoretically predicted dipole moments, if the silicon clusters are regarded as flexible particles and the nozzle temperature is used as the internal temperature (see Eq. 3.44). Interestingly, strong variations in the values of polarizabilities are observed for cluster sizes up to about 30 atoms, while for larger clusters these values are both far less fluctuating and are no longer significantly increased [11, 15, 16]. This is due to the fact that at about 30 atoms the growth behavior changes. Clusters up to this size exhibit elongated structures while for $N > 30$ a spherical growth behavior is observed. The observed effective polarizabilities show impressively that the larger, more spherical clusters do not possess pronounced dipole moments. Further cooling of the silicon clusters leads to the freezing of a dedicated structural configuration. In such a case, observed beam profiles need to be characterized via a simulation of the rotational dynamics of an, in general, asymmetric rotor [12]. Thus, in an ideal case, the structure of silicon clusters can be determined by identifying the corresponding isomer that describes the beam profiles best. By employing such an approach, it was shown that for example the (global minimum) structure of Si_8 consists of a centro-symmetric bicapped octahedron.

(c) The structures of silicon and germanium are closely related in the solid state which explains why the structural motifs of silicon and germanium clusters are often alike. Nevertheless, differences are also observed. For instance, Ge_8 exhibits a polar structure, a single capped pentagonal bipyramid, in contrast to Si_8 [12, 18]. In order to explore to what extent electric deflection methods can be applied to determine the electric dipole moment and ultimately even the geometric structure of clusters, detailed studies were undertaken for some Ge_N-clusters to assess the differences that arise from applying perturbative methods in comparison to an exact classical and quantum mechanical treatment of the rotational dynamics in the electric field, respectively [17]. Thereby, it was possible to study in more detail the deviations from the symmetry of a spherical rotor as well as the influence of finite rotational temperatures. For example, the shift and broadening of the beam profile as a function of field strength were showcased for one asymmetric isomer of Ge_9 and Ge_{10} (see Fig. 4.5). First order perturbation theory predicts independent of the structure, that the shift increases linearly, while the broadening should remain independently of the electric field strength. The exact, classical treatment of the rotational dynamics shows that the polarizability can be obtained with good accuracy by applying perturbative methods, while, depending on the structure of the examined cluster, a determination of the dipole moment is definitely not possible with perturbation theory. The limitations of the method to determine the structure arise at the moment mainly from the fact that the dipole moment distribution function cannot be reconstructed explicitly from the beam profile. For example, it should be possible to differentiate the two Ge_9 isomers with the lowest energy via their different dipole moment distribution functions. This has not been achieved experimentally, even though the orientation of the dipole moment within a molecule-fixed coordinate system is substantially different for both structures. This is because the dipole moment distribution function must be obtained by deconvoluting the beam profile. This deconvolution process is, however, ambiguous since too few measuring points are available for the beam profile, and thus reflects in essence only the magnitude of the dipole moment, which is approximately the same for both isomers. Hence, with the deflection profiles determined so far, no clear differentiation between the two Ge_9-isomers is possible (Fig. 4.5).

(d) The electric deflection behavior of small tin clusters were analyzed at low nozzle temperatures, too. By simulating the corresponding beam profiles of the energetically lowest-lying isomers it was possible to identify the structural isomers present

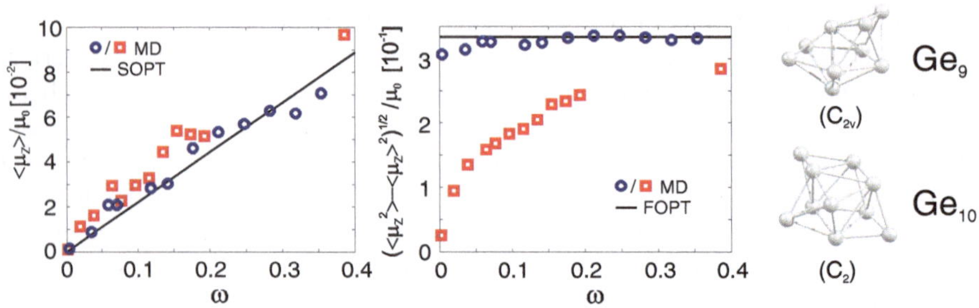

Fig. 4.5 Simulation of the deflection behavior of two asymmetric rotors, whose structures are depicted in the right part of the figure [17]. Ge_9 and Ge_{10} are severe asymmetric rotors where the dipole moment of Ge_9 is orientated along the molecular internal c-axis and that of Ge_{10} along the b-axis. The dependence of the average value of the projection of the dipole moment (*left graph*) to the laboratory z-direction and the square root of the variance of the dipole moment (*right graph*) are plotted as a function of $\omega = \mu_0 E/k_B T_{rot}$. Both quantities are normalized to the magnitude of the dipole moment. Non-perturbative classical results (Ge_9: *blue circles*, Ge_{10}: *red squares*) are compared to predictions employing spherical FOPT and SOPT. It can be seen that the average values $\langle \mu_z \rangle$ obtained from SOPT agree well with those obtained from a non-perturbative molecular dynamics (MD) simulation, i.e. same values for the polarizability are obtained. The variance $(\langle \mu_z^2 \rangle - \langle \mu_z \rangle^2)$ can also be satisfactorily predicted for Ge_9 using FOPT, but the perturbative approach breaks down in the case of Ge_{10}, where FOPT would yield completely wrong values for the dipole moment. This is due to the fact that the dipole moment is orientated along the molecular internal b-axis for Ge_{10} and that in the case of small field strengths the rotational motion of an asymmetric rotor averages out and the contribution of the b-component to the dipole moment completely vanishes. The alignment of the dipole moment in the electric field becomes stronger and stronger at higher field strength, resulting in increasing values of the normalized variance. Reprinted with permission from Heiles et al. [17]. Copyright 2011, American Institute of Physics

in the molecular beam [19]. A strong similarity between silicon, germanium and tin clusters was revealed [19, 20]. The influence of the internal temperature of the clusters was investigated in detail for tin clusters (see Fig. 4.6). The deflection behavior of Sn_{10} is now showcased as an example. The ground state structure of Sn_{10} consists of a tetra-capped trigonal prism with a theoretically predicted dipole moment of 0.6 D. The observed beam profile at low temperatures can be perfectly described by the rotational dynamics of a prolate, symmetric rotor by taking the theoretically predicted values of the inertia tensor and the molecule-fixed dipole moment into account. When the nozzle temperature and therefore also the internal temperature of the cluster is increased, the beam broadening gradually disappears due to the permanent dipole moment until at a nozzle temperature of 100 K only a one-sided shift of the beam profile can be observed. The derived effective polarizability is, however, still significantly higher than the theoretically predicted value of 7.02 Å^3. Obviously, increasing the nozzle temperature leads to a thermally stronger excited cluster whose dipole moment can fully relax in the electric field. The continued presence of an additional dipolar contribution can be quantitatively explained by a Langevin-Debye-like behavior if the internal temperature is set equal to the nozzle temperature. The interesting question is how the relaxation of the dipole moment takes place? To answer

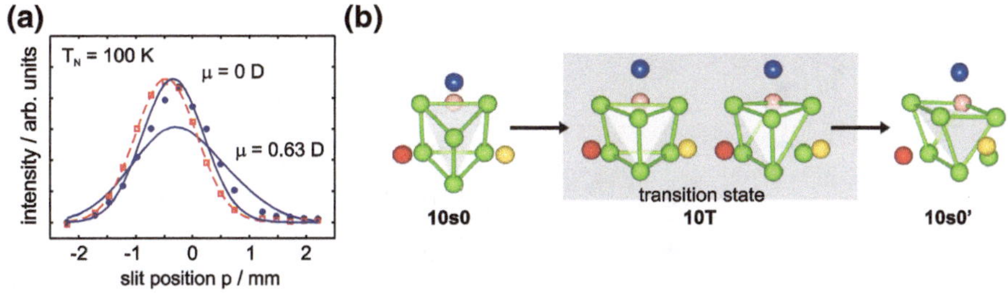

Fig. 4.6 Influence of the nozzle temperature on the deflection behavior of Sn_{10}. In **a** molecular beam profiles for Sn_{10} without (*red squares*) and with an electric field (*blue circles*) are depicted at a deflection voltage of 28 kV and a nozzle temperature of 100 K [19]. In addition, molecular dynamics simulations of a rigid symmetric rotor with the theoretically predicted value of the electronic polarizability of 7.02 $Å^3$ and a dipole moment of 0 and 0.63 D, respectively, are shown as *blue solid curves*. The broadening of the observed deflection profile can be perfectly described by a classical simulation of the rigid rotor dynamics at a nozzle temperature of 40 K [19]. At 100 K no broadening can be observed, which indicates that the cluster can no longer be treated as a rigid rotor. However, the susceptibility is increased compared to the purely electronic polarizability. Due to the additional dipolar contribution, the simulated beam profile with a dipole moment of 0 D exhibits a too small shift. By calculating the dipolar contribution with a Langevin-Debye approach (Eq. 3.44), a good agreement with experiment is obtained, when the nozzle temperature is used as the internal temperature. **b** A possible mechanism leading to the loss of the beam broadening in the case of Sn_{10} was identified by a quantum-chemical study of the potential energy hypersurface. In addition to the ground state 10s0 an almost degenerate transition state 10T ($\Delta\epsilon = 0.08$ eV) exists, so that during thermal excitation an isomerization dynamics (10s0 $\rightarrow$ 10T $\rightarrow$ 10s0′) takes place, and thus the dipole moment rotates within the laboratory-fixed coordinate system. Due to successive isomerizations the time-averaged dipole moment and the beam broadening disappear. Reprinted with permission from Schäfer et al. [19]. Copyright 2008, American Chemical Society

this question, the potential energy surface was analyzed in detail and it was found that a transition state exists which is almost degenerate to the ground state. Therefore, an isomerization dynamics is thermally instigated which resembles a pseudorotation and thus yields a corresponding fluctuating dipole moment. By estimating the frequency of this process with a simple Arrhenius-Ansatz

$$\nu = \nu_0 \cdot e^{\frac{-E_a}{k_B T}},$$

one obtains transition times of 20 ns if theoretically calculated activation energy E_a of 0.08 eV and the value for the imaginary frequency ν_0 of 500 GHz along the reaction coordinate at $T = 100$ K are considered. This is already significantly shorter than the residence time of the clusters in the electric field. Therefore, it is quite possible that the probability of the occurrence of a dipole moment orientation under the present experimental conditions are given in the statistical limit of a canonical distribution, thus justifying the Langevin-Debye approach. It is interesting in this context that, at elevated internal temperatures, the two centro-symmetric clusters $Sn_{6/7}$ have an

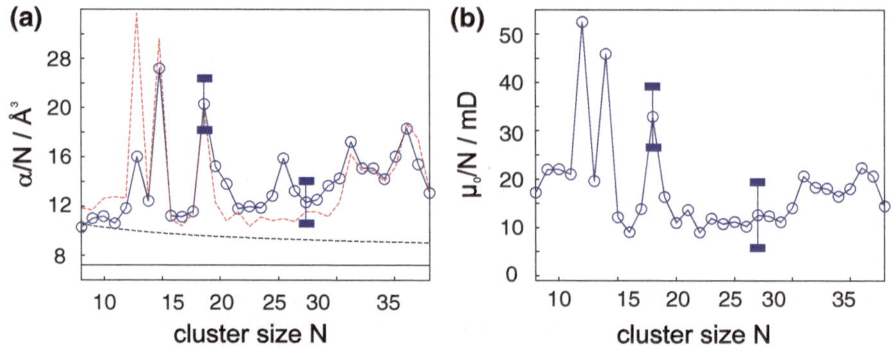

Fig. 4.7 Size dependence of the dielectric properties of Pb_N clusters [22]. **a** Effective polarizabilities per atom (*blue circles connected by blue line*) as a function of cluster size, obtained from first order perturbation theory (FOPT) from the deflection of the beam profile (see Eq. 3.21) in an electric field with deflection voltages between 10 and 28 kV at a nozzle temperature of 50 K [22]. The experimentally obtained values are significantly larger than the polarizabilility of a small metallic sphere having the properties of solid lead (*black, solid line*). This discrepancy remains even when a realistic value for the electronic "spill-out" length of 0.5 Å (*black, dashed line* [23]) is considered. Particularly striking are the large susceptibilities of $N = 12$, 14 and 18, which are accompanied by a distinct observed beam broadening and indicate the presence of a permanent dipole moment. Dipole moments per atom also obtained via FOPT from the beam broadening (see Eq. 3.18) are plotted in **b** [22]. It is striking that large values of the dipole moment correlate with corresponding anomalies of the susceptibility. Therefore, the increased effective polarizabilities can be explained at least in part by an additional dipole contribution (*red, dashed line* in **a**), by accounting for an additional adiabatic polarization of the cluster via second order perturbation theory (SOPT, see Eq. 3.27). A rotational temperature of 3 K is best suited to describe the experimental values. Reprinted with permission from Schäfer et al. [22]. Copyright 2008, American Institute of Physics

additional vibration-induced contribution to the polarizability [21]. The microscopic mechanism for this behavior is, however, not yet fully understood.

(e) The heaviest stable element of group 14 is significantly different in its behavior from the lighter homologues. Thus, it can be expected that lead clusters have other structural and dielectric properties. The accurate theoretical prediction of the cluster structure and the electric dipole moment is in these heavy element containing clusters more difficult since relativistic effects play a major role. A non-perturbative simulation of the deflection behavior is, therefore, not easily possible, so that the observed deflection profiles at a nozzle temperature of 50 K had to be evaluated within the framework of first order perturbation theory (see Fig. 4.7) [22]. It is assumed that the clusters can be described by rigid, approximately spherical tops and that the interaction energy with the electric field is small compared with the rotational energy. Effective polarizabilities determined by means of the beam shift are for almost all lead clusters substantially larger than the polarizability of a small metallic sphere with the properties of macroscopic lead. This discrepancy remains apparent, even if a realistic "spill-out" is taken into account. It is noteworthy that for some cluster sizes like $Pb_{12/14/18}$ a significant beam broadening is observed, which implies a permanent electric dipole moment. A small beam broadening can be detected for all other cluster sizes. Therefore, it seems likely that the increased effective polarizabilities are related to the permanent dipole moments. Second order perturbation theory (Eq. 3.27) can take this into account, i.e. a polarization occurs to the clusters entering the electric field adiabatically. Assuming that lead clusters can still be described as spherical, rigid rotors, the additional contribution to the polarizabilitiy can be considered by employing the values of the electric dipole moment which were obtained from the beam broadening via first order perturbation theory. In doing so, the trend of the observed polarizabilities can be described remarkably well. The obvious, existing discrepancies in the case of, for instance, Pb_{12}, Pb_{14} or Pb_{25} may have different causes: The clusters can be asymmetric or non-rigid, or the perturbative approach fails altogether. For Pb_{14} it was shown that a non-perturbative approach can perfectly describe the beam profile, if one still adheres to a rigid, spherical structure. However, this was not the case for Pb_{12} and Pb_{25}. Reliable theoretical predictions of the inertia tensor and the dipole moment would be necessary in order to conduct classical simulations of the rotational dynamics in an electric field. In addition, it must be noted that the somewhat larger lead clusters like Pb_{25} are probably not completely rigid, since weak vibrational modes are associated with heavy element containing clusters and these modes are already thermally excited at a nozzle temperature of 50 K. This would result in a smaller polarizability than expected from an adiabatic polarization model and ultimately observed in the experiment. For a quantitative understanding of the dielectric properties of lead clusters one experimental requirement beside detailed quantum-chemical studies are even lower nozzle temperatures. Nevertheless, realizing that a typical metal cluster like Pb_N does not possess shielded electric dipole moments is remarkable and thus underlines the dramatic difference between condensed matter and small clusters.

4.3 Core-Shell Clusters and Nanoalloys

In this subsection, we will show how the dielectric properties of clusters are influenced by the chemical composition. For a start we consider as an example clusters of group 14 that were doped or alloyed with a secondary metal.

(a) The addition of the electropositive metal magnesium to lead clusters is discussed first. The bimetallic species are brought into the pulsed molecular beam by means of laser vaporization of a mixed target, are then photoionized using an ionization laser (157 nm) and finally detected via a time-of-flight mass spectrometer [24]. Interestingly, lead clusters that are singly doped with magnesium are formed predominantly. The fact that the lead clusters must have a size of at least 8 or 9 atoms before singly doped clusters can be detected in the mass spectrum indicates that the magnesium atoms do not just attach to the surface of the lead cluster, but that they are placed in cages made out of lead atoms. This endohedral inclusion has been predicted theoretically, i.e. magnesium is incorporated in highly symmetric cluster structures [25]. For example, the inclusion of a Mg-atom in Pb_{12} yields an icosahedral cluster of the type $Mg@Pb_{12}$. The influence of doping on the structure of the lead cluster can be determined particularly well with the aid of electric deflection experiments. While the beam profile of Pb_{12} with its polar (i.e. non-icosahedral) structure exhibits a strong broadening in the electric field, this beam broadening disappears completely after insertion of a Mg-atom. If the magnesium atom would be incorporated on the surface or within the skeleton of the lead cluster, then the doped cluster would also be polar and should still exhibit a strong beam broadening. No or only very small dipole moments have been observed for the other singly doped clusters, which is another indication for an inclusion of the magnesium atom into a cage of lead atoms. The endohedral position of the dopant atom can be justified by a charge transfer taking place between the magnesium atom and the lead cage, so that the cluster is better described by the ionic resonance structure $Mg^{2+}@Pb_{12}^{2-}$. The corresponding doubly negatively charged lead cluster cages thus represent particularly stable Zintl species, which according to the Wade-Mingos rules are *closo*-clusters and should form corresponding *close*-cage structures [26, 27] (Fig. 4.8).

(b) A completely different behavior is obtained if, instead of an electropositive metal, the doping process is carried out with a more electronegative element, such as bismuth for instance [28, 29]. For that purpose, tin clusters doped with bismuth atoms were studied. In this case, there is only little charge transfer between the original cluster cage and the dopant atom, but instead bismuth is included into the cage of tin atoms. However, it can cause a severe reconstruction of the cage structure, as the substitution of a tin atom with bismuth changes the number of valence electrons in the cluster. Thus, the number of valence electrons can be increased sequently by successive substitution. Electric deflection experiments can therefore be utilized, to monitor the influence of the doping level (i.e. the number of valence electrons) on the dielectric properties. For this purpose, clusters with a total number of 9 atoms are considered: For the cluster species Sn_9, Sn_8Bi, Sn_7Bi_2 molecular beam broadening

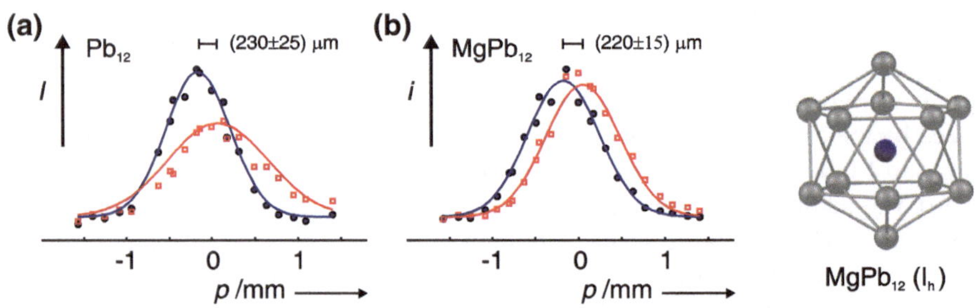

Fig. 4.8 Molecular beam deflection profiles (signal intensity *i* is plotted against the position *p* of the moveable slit) for Pb_{12} (**a**) and $MgPb_{12}$ (**b**) without (*filled circles*) and with (*open squares*) an applied electric deflection field of $1.9 \cdot 10^7$ V/m at a nozzle temperature of 50 K [24]. In the case of Pb_{12} beam broadening is obvious, indicating the presence of an electric dipole moment. For $MgPb_{12}$, on the other hand, only a one-sided deflection is evident and thus no permanent dipole moment is observed. The theoretically predicted structure of $MgPb_{12}$, shown in the *right* part of the figure, is consistent with the observed deflection experiments, as a permanent dipole moment is ruled out due to the inversion symmetry [24]. With permission from Schäfer et al. [24]. © 2008 Wiley-VCH Verlag GmbH & Co. KGaA, Weinheim

is observed in the electric field, i.e. polar aggregates are present. The exchange of another tin atom with Bi, i.e. Sn_6Bi_3, leads to the disappearance of the broadening, thus indicating a non-polar cluster. In order to explain this behavior, the ground state structures of these doped clusters were searched and optimized by means of a genetic algorithm. It is found that doping tin with bismuth yields cluster structures with a trigonal prism as a main motif, which are tricapped and completely analogue to the corresponding structures of the tin cluster anions. That is, the inclusion of bismuth transfers formally a neutral tin cluster into the cluster anion with the corresponding structure. This means that for Sn_6Bi_3 a nonpolar, centro-symmetric structure is obtained, just like it is observed in the experiment. A quantitative comparison between theory and experiment can be made via a classical simulation of the rotational dynamics in the electric field. This reveals an excellent agreement between the simulated beam profiles for the ground states and the experimental data (Fig. 4.9).

(c) Other effects result from doping fullerenes with electropositive metals of the first main group. Here, neither an endohedral inclusion of the metal atom into the carbon cage, nor a modification of the cage structure takes place, but the dopant atom is adsorbed on the cluster surface [31]. Depending on the internal temperature of the

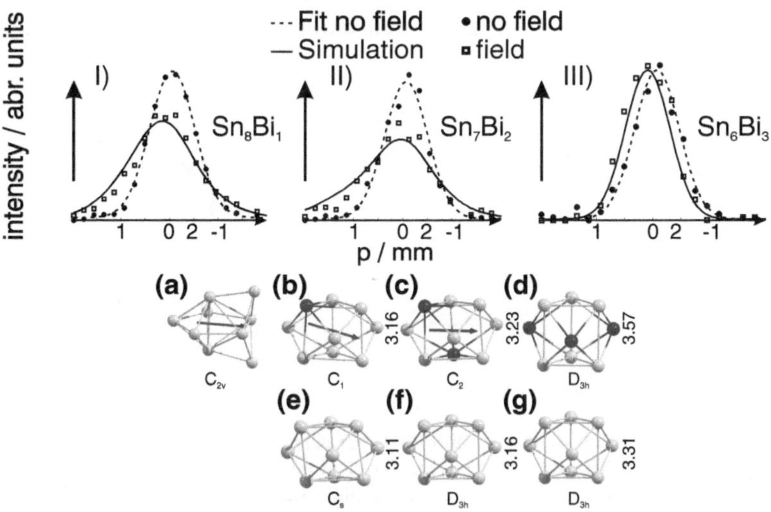

Fig. 4.9 (*Top*) Molecular beam deflection profiles (signal intensity is plotted against the position p of the variable slit) for tin clusters doped with bismuth without (*filled circles*) and with (*open squares*) an applied electric deflection voltage of 15 kV for Sn_8Bi_1 (I) and Sn_7Bi_2 (II), and 25 kV for Sn_6Bi_3 (III) at a nozzle temperature of 33 K [28]. The simulations of the deflection profiles (*solid curves*) were performed with a non-perturbative method using the theoretically predicted ground state structures at a rotational temperature of 6 K (*bottom*). The calculated structures of (**a**) Sn_9, (**b**) Sn_8Bi_1, (**c**) Sn_7Bi_2, (**d**) Sn_6Bi_3, (Sn *gray*, Bi *blue*) are depicted together with their molecular point groups on the *bottom* part of the figure, and are compared to the corresponding structures of the valence-isoelectronic anions (**e**) Sn_9^-, (**f**) Sn_9^{2-}, (**g**) Sn_9^{3-} [28]. The *blue arrows* show the electric dipole moments, the numbers indicate the bond lengths in Å along the (pseudo)-C_3-axis. With permission from Heiles et al. [28]. © 2012 Wiley-VCH Verlag GmbH & Co. KGaA, Weinheim

cluster, the metal atom is during its residence time in the electric field either localized or mobile. The clusters were synthesized using a dual-laser vaporization source. Deflection profiles were measured after photoionization with a position-sensitive time-of-flight mass spectrometer. For example, in low-temperature experiments a strong broadening of the beam profile was observed for NaC_{60}, while in the high temperature limit only a one-sided shift was observed (see Fig. 4.10) [32]. The strong broadening at low temperatures is due to the large dipole moment of the doped cluster of 13.93 D, which was determined by quantum-chemical calculations and can be traced back to an almost complete charge transfer between the sodium atom and the C_{60} cage. The observed beam profile can be perfectly reproduced by means of a classical simulation when the symmetric structure of NaC_{60} and its theoretically predicted dipole moment are taken into account. The adsorbed sodium atom becomes increasingly mobile by rising the temperature, and the deflection profiles, hence, can no longer be described by a rigid rotor model. The additional dipolar contribution to the effective polarizability which is still present can be described within a classical model considering the thermally activated hopping movement of Na on the C_{60} surface. The dipole moment and, in addition, the relaxation time for the hopping dynamics were determined via adjusting to the observed beam profiles. The dipole moment is in good agreement with the value that resulted from the simulations of the low-temperature experiments. The relaxation time exhibits an Arrhenius-like

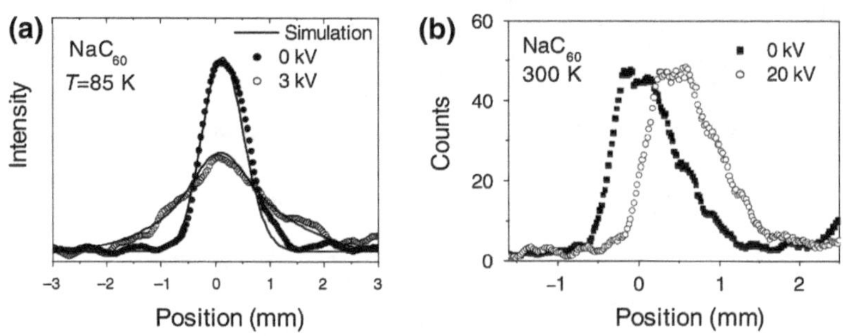

Fig. 4.10 Molecular beam deflection profiles for NaC_{60} at two different nozzle temperatures of 85 K (**a**) and 300 K (**b**), respectively [5]. The data points were obtained without (*filled circles and squares*) and with (*open squares*) an applied deflection voltage of 3 kV (85 K) and 20 kV (300 K), respectively. The low-temperature experiments exhibit a significant broadening of the beam profile, although only a relatively low deflection voltage was applied. This indicates the presence of a large permanent dipole moment. At room temperature, only a one-sided shift of the beam profile is observed, and this can be explained by a fluctuating dipole moment. The beam profiles at 85 K were simulated using a rigid, symmetric rotor with a dipole moment of 14.8 D (*solid curves*), where the limiting case of small electric fields and high rotational temperatures are assumed ($\omega = \mu E / k_B T_{rot} \rightarrow 0$ and $1/T_{rot} \rightarrow 0$). The value of the dipole moment used is in good agreement with quantum-chemical predictions (B3LYP [7] with LANL2DZ [30] and SDD-Basis [8]). The room temperature behavior can be explained by a rapid movement of the sodium atom on the surface of C_{60}. This results in an additional dipolar contribution to the susceptibility which can be correctly predicted by means of a Langevin-Debye approach with an internal temperature of 300 K. Reprinted with permission from Broyer et al. [5] Copyright 2007, Institute of Physics

behavior, where the activation energy for the hopping process is a mere 0.02 eV. This confirms the already known behavior that alkali metal atoms are highly mobile on the graphite surface, even at room temperature. The dipole moments of other clusters MC_{60} with M = Li, K, Rb and Cs were also determined [32]. It is found that the dipole moment increases with increasing atomic number of the alkali metal, indicating a greater distance between the positively charged alkali metal and the negatively charged C_{60} cage (Fig. 4.10).

(d) Taking a step further, the question of what would happen if more than one sodium atom is adsorbed on the fullerene surface comes to mind. Are the carbon atoms going to be coated by the sodium atoms or are the metal atoms going to form their own cluster on the fullerene surface? Theoretical studies suggest that up to 8 Na atoms can adsorb onto the surface, but that thereafter sodium cluster formation occurs on the fullerenes [33]. Deflection profiles were observed experimentally for Na_NC_{60} with $n = 1$–34, which exhibit no broadening but a significant one-sided shift. Values for the effective polarizability obtained from the broadening are in the region of 700 and 2500 Å^3, and are well above the polarizability of C_{60} (76.5 Å^3) and the isolated Na_N clusters ($\approx N \cdot 16$ Å^3) (see Fig. 4.11) [5, 34]. The enormously enlarged polarizabilities indicate the presence of a large electric dipole moment,

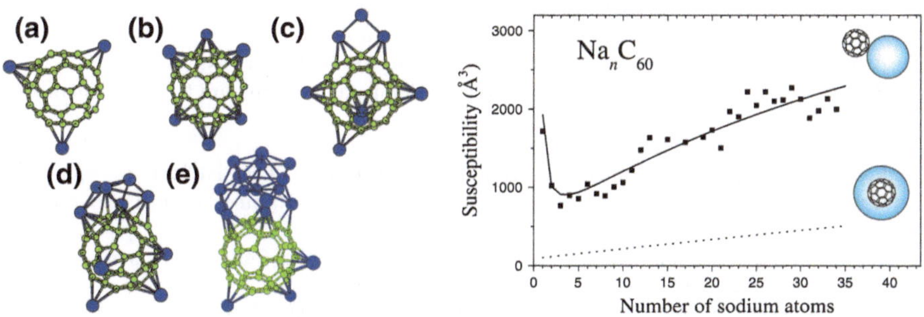

Fig. 4.11 The *left* part of the figure depicts quantum-chemically optimized ground state structures of Na_NC_{60} clusters with $N = 3$ (**a**), 6 (**b**), 8 (**c**), 10 (**d**) and 20 (**e**) atoms (Na—*big blue*; C—*small green*) [5, 33]. It is evident that after a certain number of sodium atoms the fullerene cage is not coated uniformly any more, but that sodium clusters are formed on the C_{60} surface. The *right* part illustrates susceptibilities obtained from room temperature deflection experiments for $N = 1$–34 (*filled squares*) [5, 34]. Only a one-sided shift of the molecular beam is observed in the electric deflection profiles. The resulting values of the effective polarizability are significantly larger than for a single C_{60} molecule plus an isolated Na_N cluster. This is due to the fact that a charge transfer takes place between the electropositive Na atom and the fullerene. This results in clusters that possess a permanent dipole moment which fluctuates strongly at room temperature. Thus, a very large dipolar contribution to the susceptibility arises. The observed size dependence of the susceptibility is in agreement with the theoretically predicted growth mechanism that predicts the formation of sodium clusters on the C_{60} surface (*solid line*). The change of the effective polarizability as a function of the number of Na atoms, if sodium were to coat the carbon surface uniformly, is also denoted (*dashed line*) (*right*). Reprintred with authors permission. We also thank F. Calvo for providing the colored cluster structures (*left*). Reprinted with permission from Broyer et al. [5]. Copyright 2007, Institute of Physics

which can statistically align with the electric field, thus leading to an additional dipolar contribution to the observed electric susceptibility. Hence, the observed values of the susceptibility eliminate the possibility of a uniform coating of the fullerene, as such structures would exhibit no or only a weak dipole moment and one would therefore not observe such strongly increased values for the effective polarizability. A structural model where the fullerene is at first coated with Na atoms, reducing the effective polarizability, is in accordance with the observed trend of the susceptibility. From a size of about 8 atoms, sodium clusters are formed which donate electrons to the fullerene cage, thus giving rise to a large dipole moment. This dipole moment can be estimated from the geometric structure of the sodium-fullerene composite in order to calculate the dipolar contribution to the effective polarizability. This results in a very good agreement between the model and experimental values, as shown in Fig. 4.11.

References

1. Compagnon I, Antoine R, Rayane D, Broyer M, Dugourd P (2002) Phys Rev Lett 89:253001
2. Moro R, Bulthuis J, Heinrich J, Kresin VV (2007) Phys Rev A 75:013415
3. Moro R, Rabinovitch R, Xia C, Kresin VV (2006) Phys Rev Lett 97:123401
4. Carrera A, Mobbili M, Marceca E (2009) J Phys Chem A 113:2711
5. Broyer M, Antoine R, Compagnon I, Rayane D, Dugourd P (2007) Phys Scr 76:C135
6. Rayane D, Allouche AR, Antoine R, Broyer M, Compagnon I, Dugourd P (2003) Chem Phys Lett 375:506
7. Becke AD (1993) J Chem Phys 98:5648
8. Fuentealba P, Preuss H, Stoll H, Von Szentply L (1982) Chem Phys Lett 89:418
9. Compagnon I, Antoine R, Broyer M, Dugourd P, Lermé J, Rayane D (2001) Phys Rev A 64:025201
10. Ashcroft NW, Mermin DN (2007) Festkörperphysik. Oldenburg Verlag, München
11. Götz D, Heiles S, Schäfer R (2012) Eur Phys J D 66:293
12. Götz DA, Heiles S, Johnston RL, Schäfer R (2012) J Chem Phys 136:186101
13. Deng K, Yang J, Chan CT (2000) Phys Rev A 61:025201
14. Jackson KA, Yang M, Chaudhuri I, Frauenheim T (2005) Phys Rev A 71:033205
15. Schäfer R, Schlecht S, Woenckhaus J, Becker JA (1996) Phys Rev Lett 76:471
16. Becker JA (1997) Angew Chem Int Ed 36:1390
17. Heiles S, Schäfer S, Schäfer R (2011) J Chem Phys 135:034303
18. Schäfer S, Schäfer R (2008) Phys Rev B 77:205211
19. Schäfer S, Assadollahzadeh B, Mehring M, Schwerdtfeger P, Schäfer R (2008) J Phys Chem A 112:12312
20. Shvartsburg AA, Jarrold MF (2000) Chem Phys Lett 317:615
21. Kast SM, Schäfer S, Schäfer R (2012) J Chem Phys 136:134320
22. Schäfer S, Heiles S, Becker JA, Schäfer R (2008) J Chem Phys 129:044304
23. Snider DR, Sorbello RS (1983) Phys Rev B 28:5702
24. Schäfer S, Schäfer R (2008) ChemPhysChem 9:1925
25. Rajesh C, Majumder C (2008) J Chem Phys 128:024308
26. Wade K (1976) Adv Inorg Chem Radiochem 18:1
27. Mingos DMP (1972) Nat Phys Sci 236:99
28. Heiles S, Hofmann K, Johnston RL, Schäfer R (2012) ChemPlusChem 77:532
29. Heiles S, Johnston RL, Schäfer R (2012) J Phys Chem A 116:7756

30. Wadt WR, Hay PJ (1985) J Chem Phys 82:284
31. Rayane D, Antoine R, Dugourd P, Benichou E, Allouche AR, Aubert-Frecon M, Broyer M (2000) Phys Rev Lett 84:1962
32. Rayane D, Allouche A, Antoine R, Compagnon I, Broyer M, Dugourd P (2003) Eur Phys J D 24:9
33. Roques J, Calvo F, Spiegelman F, Mijoule C (2003) Phys Rev Lett 90:075505
34. Dugourd P, Antoine R, Rayane D, Compagnon I, Broyer M (2001) J Chem Phys 114:1970

Chapter 5
Novel Experimental Tools

In the first chapters the experimental (Chap. 2) and theoretical (Chap. 3) essentials of the electric field deflection methods were discussed. The improvements of cluster sources, detectors and acquisition systems combined with quantum chemistry and the more realistic modeling of the beam deflection results have given detailed insight of numerous cluster systems as it was showcased in the previous chapter (Chap. 4). Despite the described improvements and outstanding opportunities of the electric field deflection method, several limitations exist. Due to the static electric field used in all investigations, no charged cluster can be investigated, since they would be extracted from the molecular beam. This, furthermore, does not allow to individually probe the frequency-dependent response of the polarizability and electric dipole moment, since both quantities will be studied simultaneously with a static electric field. These shortcomings can not be resolved by further improvements of the experimental setup but a completely new experimental approach needs to be introduced. Beside this interest in experimental methods for basic research, there is more to the investigation of the dielectric properties. Nearly all basic but also applied methods in the gas phase, in particular, mass spectrometry and spectroscopy, investigate ions. The trajectories of ions can be manipulated easily by electric or magnetic fields. On the other hand, many interesting substances and molecules are neutral and introducing a charge in order to study these intrinsic properties could alter the outcome of the experiments. Therefore, it is highly desirable to be able to manipulate the movement of neutral objects in a similar way as done for ions.

Based on these ideas, the upcoming section will contain the description of newly developed experimental tools, which can overcome some of the intrinsic limitations of the electric field deflection method. These methods use static and varying electric fields, enabling the experimental determination of dynamic polarizabilities (Sect. 5.1) or the manipulation of the motion of neutral species in the desired way (Sect. 5.2). This description does not intend to be complete but rather to introduce the basic experimental and theoretical principles using different instructive examples.

S. Heiles and R. Schäfer, *Dielectric Properties of Isolated Clusters*, SpringerBriefs in Electrical and Magnetic Properties of Atoms, Molecules, and Clusters,
DOI: 10.1007/978-94-007-7866-5_5,

5.1 Light Force and Near Field Interferometry

As mentioned in the introductory part of this chapter the usage of a static electric field gives rise to various experimental limitations like the inability to investigate charged clusters, the fact that the influence of the permanent can not be separated from the induced dipole moment and the insensitivity for any frequency-dependent dielectric information. The at first sight quite obvious solution of the problem is to use an AC electric field $\mathbf{E}(\mathbf{r}, t)$. But what frequency of the AC field should be used? This question is easily answered if one takes the processes, which need to be eliminated, into account. First, the experiment should be able to discriminate between the permanent and the induced electric dipole moment. Since $\boldsymbol{\mu}_0$ for a rigid compound is coupled to the atomic framework and this itself undergoes a rotational motion in the gas phase the frequency of the AC field should be larger than the rotational frequency of the cluster. Thus, with using frequencies larger than $\sim 10^{11}$ Hz, the clusters and hence $\boldsymbol{\mu}_0$ can not adapt to the rapidly oscillating electric field strength. Second, all kind of photo-physical processes like absorption or ionization, which are resonance enhanced, should be omitted in order to mainly detect the dielectric response of the polarizability to the electric field. From these pre-requirements it can be concluded that frequencies in the range of 10^{13}–10^{15} Hz are most promising to undertake this experiment. Consequently, this leaves only light as a possibly source of the electric AC field. The light needs to be coherent and additionally periodic since otherwise the effect of the oscillating electric field would cancel. Based on these ideas Nairz et al. designed the experimental setup schematically shown in Fig. 5.1 [1]. The layout of the molecular beam apparatus is very similar to classic experiments of Knight [2] or Kresin [3]. A beam of clusters, in this case C_{60} and C_{70}, is produced in an oven, the clusters are velocity selected, collimated and detected after passing a region of free flight. The innovative part of the setup is replacing the deflection electrodes by a continuous Ar^+-Laser system together with some optical components. The up to 27 W from the laser with various lines in the visible light region are split up in two parts by dichroic mirrors. In order to ionize the clusters for detection, multiple colors with 17 W are used. The second part of the laser beam used in the experiment contains only light of $\lambda_L = 514.5$ nm and its intensity P_0 can be varied continuously between 0 and 9.5 W by rotating a $\lambda/2$-plate (Fig. 5.1). This laser beam is focused through a cylindrical lens on a plane mirror and the incoming and reflected laser beam form a standing light wave with a period of $\lambda_L/2$. What will happen with clusters that interact with this standing wave? The experimental results for C_{60} using different laser powers are shown in Fig. 5.1. It is clearly seen that no shift and no simple broadening is detected but depending on P_0 a periodic deflection pattern is observed. In order to qualitatively understand this observation we will assume that no absorption of photons takes place. Hence, in a very simple picture only the ground state electron density will respond to the electric field. The interaction potential between the cluster and the standing laser wave can be expressed by an induced dipole potential of the form

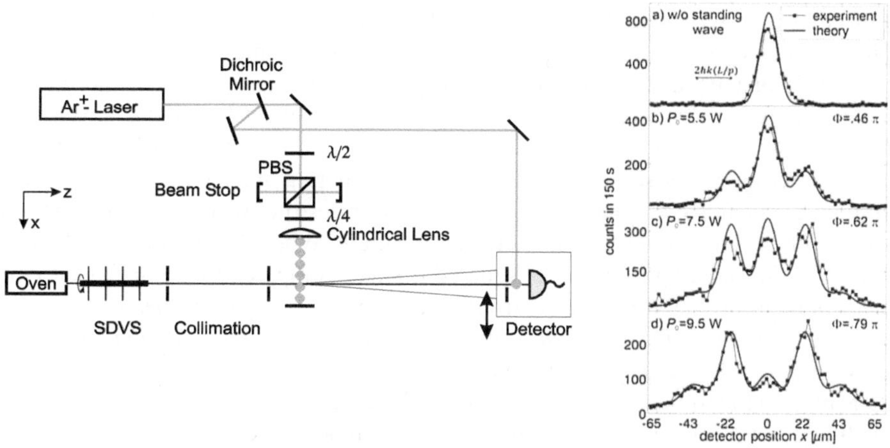

Fig. 5.1 (Schematic experimental setup, *left*) C_{60} and C_{70} clusters are expanded from an oven forming a molecular beam. After a slotted disk velocity selector (SDVS) stage and collimation the clusters interact with a standing light wave (514.5 nm, with the corresponding wave vector k_L) of variable power. Due to the high laser power used in the experiment a polarizing beam splitter (PBS) is required to prevent laser light to be back-reflected. About $L = 1.2$ m behind the interaction region the fullerenes are ionized and detected by scanning the detector over the molecular beam in 2 μm steps. (Experimental results, *right*) The periodically deflected molecular beam profiles of C_{60} (*black squares*) with the corresponding momentum in the z direction, called p, show a drastic influence of the laser power P_0 (the first profile is recorded without [w/o] applied deflection laser). A simulation (*solid line*) taking the polarization forces and absorption processes into account gives an excellent agreement with the experimental observations [1]. Reprinted figure with permission from Nairz et al. [1]. Copyright 2001 by the American Physical Society

$$V = -\frac{1}{2}\alpha(\lambda_L)\langle E(x,y,z,t)^2\rangle = -\frac{\alpha(\lambda_L)I(x,y,z)}{2c\epsilon_0}. \tag{5.1}$$

Here, $\alpha(\lambda_L)$ is the electronic ground state polarizability for λ_L, $\langle E(x,y,z,t)^2\rangle$ is the time averaged square magnitude of the electric field, $I(x,y,z)$ is the local laser intensity, c is the speed of light and the axis x, y and z are defined in Fig. 5.1. Arndt and co-workers [1] used an elliptic Gaussian laser profile of the form

$$I(x,y,z) = \frac{8P_0}{\pi w_y w_z}\exp\left(-\frac{2y^2}{w_y^2} - \frac{2z^2}{w_z^2}\right)\cos^2\left(\frac{2\pi}{\lambda_L}x\right) \tag{5.2}$$

and assumed that the clusters travel along $y = 0$ what is a very good approximation when the molecular beam is collimated adequately in the y-direction. In Eq. 5.2 the laser power enters as P_0 and the width of the profile in y and z direction is given by w_y and w_z, respectively. Having an expression for the potential and the laser intensity, the force and, consequently, the deflection easily can be derived as it was shown in Chap. 2 for classical beam deflection experiments. But for what follows it is much more instructive to look at the problem from another perspective. The clusters can be viewed as propagating matter waves, which interact with a periodic light structure. This, of course, is completely analogue to classic interferometry experiments [4, 5] but instead of a mechanical an optical grating is used. In interferometry experiments, the incoming matter waves, which are characterized by their de Broglie wavelength λ_{dB} (for all described experiments $\lambda_{\mathrm{dB}} \approx (1-5)$ pm), diffract from the grating and each scattering center serves as a source of new spherical matter waves. Behind the grating, the matter wave interfere constructively or destructively depending on their relative phase $\Phi(x)$ and form an interference pattern. When the aperture width (here related to the standing light wave) squared is small compared to the distance to the detector plane (1.2 m) times the de Broglie wavelength of the particle beam, the waves have stopped to interfere with each other before they are detected. Hence, the waves arriving at the detection plane are plane waves which have been modulated by the transmission function $t(x) = \exp(-i\Phi(x))$ (far field approximation). Consequently, the function $t(x)$ which describes the transmission properties of the grating and the corresponding phase $\Phi(x)$, are the only things needed in order to rationalize the experiment. Generally speaking, the phase change of the matter wave must depend on the local field experienced by the cluster at position x and this effect needs to be integrated over the interaction length $z = v_z \cdot t$, where v_z is the mean cluster velocity and t is the duration of the interaction (employing the eikonal approximation). By taking into account that t is much bigger than λ_{L}/c and using Eq. 5.1 as well as Eq. 5.2, the expression

$$\Phi = -\frac{1}{\hbar}\int_{-\infty}^{\infty} V(x, 0, v_z \cdot t)dt = -\Phi_0 \cos^2\left(\frac{2\pi}{\lambda_{\mathrm{L}}}x\right) \tag{5.3}$$

with the maximum phase shift

$$\Phi_0 = 2\sqrt{\frac{2}{\pi}}\frac{\alpha}{\hbar\epsilon_0 c}\frac{P_0}{w_y v_z} \tag{5.4}$$

is obtained. It is important to note, that based on an expansion of Eq. 5.3 up to first order, the original beam profile will be modulated periodically. This is in qualitative agreement with the observation made in Fig. 5.1. Furthermore, from Eq. 5.4 it becomes clear, that the profile will change with the laser power and depend on the polarizability of the cluster. This is in tune with experiment. A quantitative simulation of the experiment can be performed if absorption processes and the free propagation of the clusters behind the light structure are considered [1]. This was done by Nairz et al. (Fig. 5.1, simulation shown as solid line [1]) and an overall excellent agreement between theory and experiment has been achieved. Consequently, these experiments allow to measure the polarizability $\alpha(\lambda_{\mathrm{L}})$ if all other quantities are determined precisely. This was not only done by Arndt and co-workers but the AC polarizabilities of Rb [6], U [7] and C_{60} [8] were determined for $\lambda_{\mathrm{L}} = 1064$ nm, too. The described method enables to determine the AC polarizability and can in general be used to study ionic clusters. On the other hand, for more complex systems only small particle densities are experimentally realized and, hence, a larger velocity distribution must be used in experiment. For these systems, a single light grating is not sufficient in order to study the dielectric properties in the described way.

In order to overcome this limitation Gerlich et al. used a so called Kapitza-Dirac-Talbot-Lau interferometer (KDTLI) [9, 12]. The experimental setup of the interferometer is schematically depicted in Fig. 5.2 [9]. Two new elements have been introduced in front and behind the standing light wave. These SiN gratings with a period of 266.38 nm are separated 105 mm from the light structure. In this experiment, all particles have to pass the optical and the two mechanical gratings before they are detected. What is the difference in this compared to the previously described experiment? Obviously, the clusters will not only diffract from the light but also from the mechanical gratings. Since the length of free flight is much smaller than in the previously described case, the waves emerging from each scattering center have not stopped to interfere. Therefore, the far field approximation used up to this point is not valid and the so called near field or Fresnel diffraction needs to be considered. Thus, the incoherent beam of molecules will diffract from the first mechanical grating. From each of its slits spherical matter waves will emerge which will propagate towards the light structure and will again be diffracted. For this kind of two grating interferometer Talbot and Laue have made a very important observation. At defined

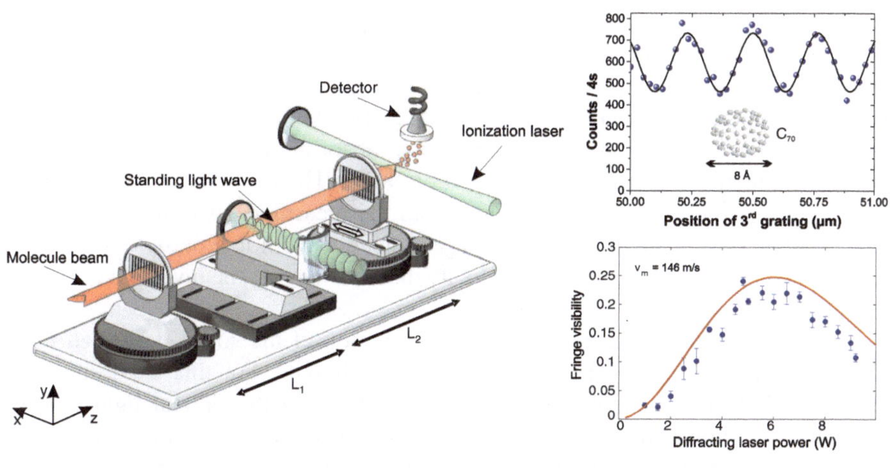

Fig. 5.2 (*Left*) Experimental setup of the KDTLI interferometer [9]. The different gratings and the standing light wave mirror can be rotated and translated in order to fine adjust the interferometer. In a typical experiment only the third grating is translated along the x-axis. The clusters passing all three gratings are ionized by 18 W of 532 nm laser light and are detected subsequently. (*Upper right*) Typical near field interferogram of C_{70} traveling with $v_z = 146$ m/s ($\pm 16\,\%$) and $P_0 = 6$ W. The *black* line is a sinusoidal fit to the recorded blue data points. (*Lower right*) Fringe visibility (Eq. 5.5) for several laser powers (*blue dots*) for C_{70}. The experiment can be reproduced if the dipole force due to the standing light wave and absorption processes are taken into account (*red curve*). We are thankful to Prof. Markus Arndt (Vienna Center for Science and Technology, Vienna) for providing the shown figures

distances from the second grating (but not too far away from it) the diffraction pattern images the intensity profile directly behind the second grating. This phenomenon was called Talbot-Laue effect and originates from the interference between different matter waves in the near field [13, 14]. The typical length for which this phenomenon is observed is called (integer) Talbot length L_T. Different waves, which emerge from two different points separated by the grating period d, will interfere constructively if the path difference is equal to λ_{dB}. Thus, a self imaging of the grating will be observed at $L_T = d^2/\lambda_{dB}$. The experiment, schematically shown in Fig. 5.2, uses this effect by setting the distance between the gratings equal to L_T resulting in a self imaging of the light grating periodicity onto the second mechanical grating. This grating is used as a in x movable aperture which let parts of the diffracted molecules reach the detection region. By sequential movement of this last interferometry element the intensity for various x positions can be used in order to reconstruct the interference pattern. A typical near field interferogram for C_{70} is shown in Fig. 5.2. As expected a sinusoidal intensity profile is observed which is a magnified self image of the standing light wave. In order to come back to the original problem of measuring the optical polarizability the influence of the laser power on the fringe visibility

$$V = \frac{I_{\text{max}} - I_{\text{min}}}{I_{\text{max}} + I_{\text{min}}} \tag{5.5}$$

needs to be evaluated [15]. Here, I_{max} and I_{min} are the maximum and minimum intensity of the interferogram, respectively, which are extracted from fitting a sinusoidal function to the data points (red curve in Fig. 5.3). The results for power dependent measurements are shown in Fig. 5.2. By taking the phase shift due to the optical grating (Eqs. 5.3 and 5.4) and possible absorption processes into account, the experimental visibility values can be reproduced. Hence, this experiment can be used to extract the optical polarizability from the measured visibility curves. This approach has the advantage that very large molecules with low particle fluxes, low degree of coherence and acceptable velocity distributions can be investigated, which is not possible by other methods [9].

A last modification is combining the information of the so far described interferometry with the classic beam deflection experiments. For this experiment, an additional deflection electrode is introduced in front of the optical light grating, schematically shown in Fig. 5.3. When the particles travel on the molecular beam axis and pass the three gratings a sinusoidal intensity pattern is detected. This is shown in Fig. 5.3a for C_{60} and C_{70}. Depending on the initial velocity distribution, different fringe visibility values are observed for the two clusters. When a voltage of 14 kV is applied to the newly introduced electrodes, the inhomogeneous electric field will give rise to an additional force $F_x = -dV_x/dx$ which of course is identical to the force introduced in Chap. 2. Consequently, this will result in an deflection of the traveling matter waves. In the case of floppy clusters with isomers possessing non-zero dipole moments (see for example [16]) this will give rise to a single sided deflection of all diffracted waves and in consequence to a deflection of the whole interference pattern. This is shown in Fig. 5.3b for the investigated fullerenes. Hence,

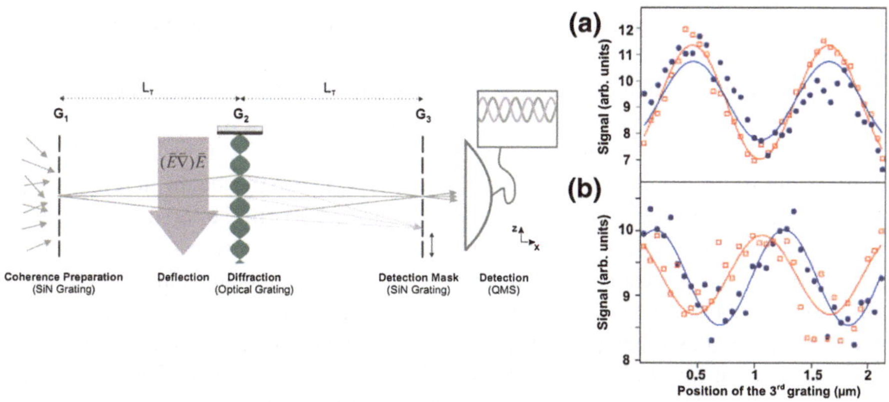

Fig. 5.3 (*Left*) Setup of the combined KDTLI/beam deflection experiment [10]. Similar to the setup shown in Fig. 5.2, the particles have to pass three gratings before they are detected. In contrast to this experiment, electrodes are placed in front of the light grating. By applying a deflection voltage to this electrodes the whole interference pattern is shifted. Reprinted with permission from Eibenberger et al. [10]. Copyright 2011, Institute of Physics. (*Right*) The interferences patterns without (**a**) and with (**b**) deflection voltage (14 kV) for C_{60} (*blue circles*) and C_{70} (*red squares*) show distinct fringe shifts [11]. Reprinted with permission from Ulbricht et al. [11]. Copyright 2008, Institute of Physics

this method reintroduces the influence of the permanent electric dipole moment as an experimental parameter but without loosing the benefits of the interference technique. If only light would be used to probe the dielectric properties of the clusters, μ_0 can not respond to the fast oscillating motion of the electric field, even if it would rapidly fluctuate. On the other hand, the shift of the matter waves caused by the permanent electric field depend on μ_0. Consequently, from the fringe shift the effective, static polarizability and from the fringe visibility the optical polarizability can be deduced [10]. This can be done for various complex molecules or clusters [10, 16] but so far no rigid cluster with a permanent dipole moment was measured employing this technique.

5.2 Stark-Modulation of Neutral Molecule Trajectories

In the last section of this chapter we will not introduce a new technique to measure the dielectric properties of an isolated molecule or cluster but rather methods that make use of these precisely determined properties and the discussed methodologies in order to modulate the trajectories of neutral molecules. The ultimate goal of these methods is to manipulate the trajectories of neutral molecules in a somewhat similar fashion as for charged particles and store them, for example, in a neutral molecule trap. These techniques then allow to study neutral molecules at very low temperatures or defined collision conditions giving deep insight in low temperature physics and chemistry. Here, we want to concentrate on the most basic of these techniques, the so called Stark deceleration, and only briefly mention other possible manipulation experiments.

A possible experimental realization of a Stark deceleration apparatus is shown in Fig. 5.4a [17]. In a rare gas seeded molecular beam of HNO_3, neutral OH molecules in the $X\,^2\Pi_{3/2}$ electronic state are formed by a ArF-Excimer Laser (193 nm) induced photodissociation of HNO_3. The carrier gas of the beam determines the average translation velocity[1] of the OH radicals while the supersonic expansion cools the rotational and vibrational degrees of freedom resulting in a predominant population of the lowest rotational state. This molecular beam enters a hexapole unit and an electric hexapole field is applied (not applied) by a high voltage switch when the molecules enter (leave) the electrodes. Since for a diatomic molecule K is not defined, at first only a second order Stark effect is expected (Eq. 3.19). However, since the radical possesses a finite orbital and electronic angular momentum, the total angular momentum $\mathbf{N} = \mathbf{J} + \mathbf{L} + \mathbf{S}$ has to be taken into account. The $\mathbf{J}$ component is the so far exclusively considered rotational contribution to the angular momentum, while $\mathbf{L}$ and $\mathbf{S}$ result from the finite orbital and electronic angular momentum. The molecular energy levels of the OH radical can be treated within the Hund coupling case (a)[2] which means that the energy levels with different orbital momentum and resulting from spin-orbit coupling are well separated. Additionally, $\mathbf{L}$ and $\mathbf{S}$ are strongly coupled to the symmetry axis of the molecule and the projection of $\mathbf{L} + \mathbf{S}$ on this axis is characterized by the quantum number Ω. Hence, for this special case, the quantum number Ω plays the same role as K does for closed-shell polyatomic molecules. During the supersonic expansion the OH radicals are cooled to the rotational and vibrational ground state but due to the $^2\Pi_{3/2}$ electronic ground state N is still 3/2 (which in this special case is equal to $|\Omega|$). Therefore, depending on the quantum number M a first order Stark effect (Eq. 3.19) can still be observed. As described in [21] this will result in a focusing of the low-field seeking (those states with increasing energy in the electric field and are consequently accelerated towards lower fields, Eq. 3.3) but a defocussing of the high-field seeking states in the hexapole deflection unit (see Fig. 5.4). In this particular case that means that only $|\Omega| = 3/2$ and $|M| = 1/2, 3/2$ with opposite signs of Ω and M (f-parity states)[3] are low-field seeking states, which will be focused and relevant for the experiment. As shown in Fig. 5.4a, at about 17 mm behind the hexapole the 102 equidistantly spaced deceleration stages with a center-to-center distance of $L = 11$ mm are placed in the molecular beam apparatus. The distance between the rods is 6 mm while the orientation of the stages alternate by 90°. With the help of fast high voltage switches ±20 kV are applied to the opposing electrodes, creating a electric field strength of $1.15 \cdot 10^7$ V/m. The radicals are subsequently state-selectively detected by laser induced fluorescence (detection wavelength 282 nm, laser beam orthogonal to molecular beam) [22]. From these measurements the population of the different low-field seeking states can be inferred. The velocity distribution is extracted from the

---

1 For example the use of Kr and Xe result in a velocity of 450 m/s and 360 m/s, respectively [17].

2 For more details see [18–20].

3 See [20] for a discussion parity labels of diatomic molecules.

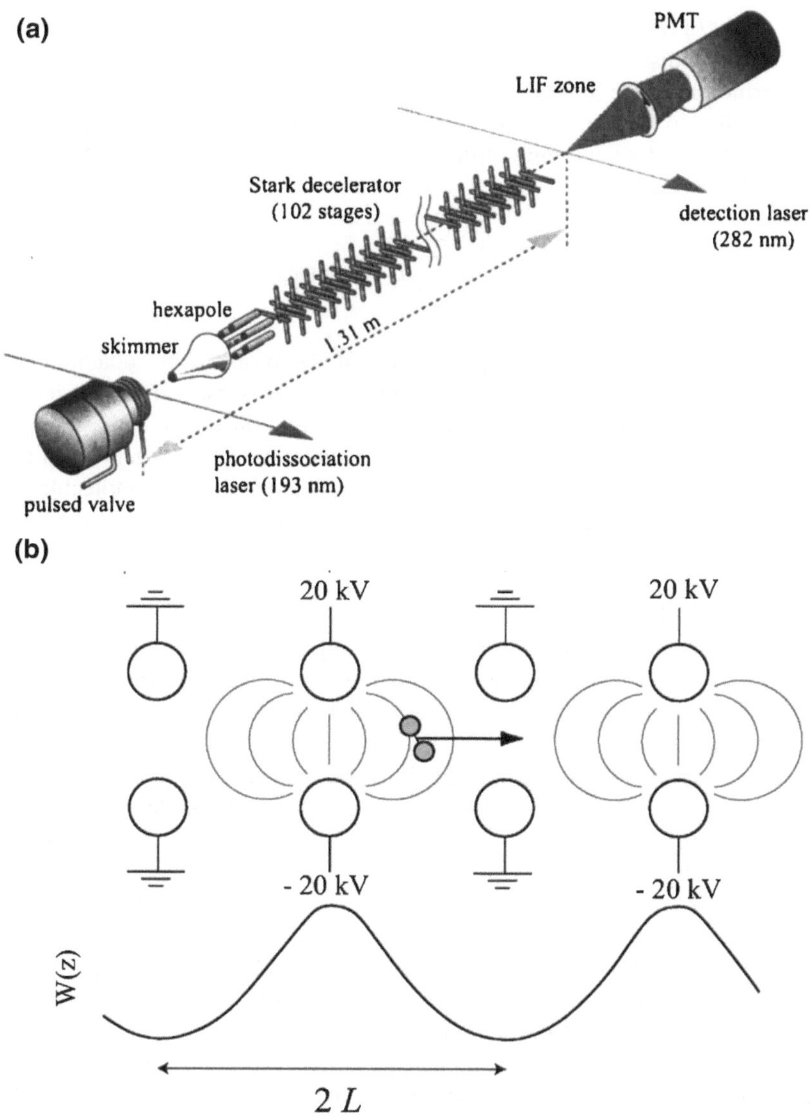

Fig. 5.4 Schematic setup of the molecular beam apparatus developed for Stark deceleration of OH radicals [23]. **a** The OH radicals are generated by an ArF-Excimer Laser (193 nm), expanded through a skimmer and the low-field seeking states are focused in a hexapole lens. After the hexapole, the Stark deceleration unit is passed and the time-of-flight (TOF) distribution is state selectively measured with laser induced fluorescence (LIF). The collected photons are detected in a photo multiplier tube (PMT). **b** In the Stark decelerator every even (odd) electrode is switched on (off) creating a periodic Stark potential energy surface $W(z)$ for a low-field seeking state with periodicity of $2L$. The maximum voltage difference is 40 kV. After every period ΔT the situation is changed and a voltage is applied to the previously grounded electrodes and the other electrodes are switched to ground and *vice versa*. Reprinted figure with permission from van de Meerakker et al. [23]. Copyright 2005 by the American Physical Society

time-of-flight (TOF) profiles, recorded by scanning the timing of the detection relative to the photodissociation laser.

The principle of the Stark deceleration experiment is illustrated in Fig. 5.4b. The voltage applied to the opposing electrodes is alternated between ground and ±20 kV (the higher the voltage the more efficient the deceleration will be) creating a Stark potential described by

$$W(\pi z/L) = \frac{a_0}{2} + \sum_{n=1}^{\infty} a_n \cos\left(n(\pi z/L + \pi/2)\right) \quad (5.6)$$

with periodicity $2L$. Due to the periodicity of the problem the potential can be described by a Fourier series given in Eq. 5.6 with the corresponding Fourier coefficients a_n. For convenience the phase $\varphi = \pi z/L$ is introduced which has the periodicity 2π (z is the axis of the longitudinal molecule motion) and can be used instead of the position to describe the motion of the molecules. When a molecule travels through an electrode configuration with the static electric fields as shown in Fig. 5.4b the net deceleration effect will be zero. For example we will assume that the shown OH radical populates a low-field seeking state. At the position shown in Fig. 5.4b the molecule will gain some potential (or Stark) energy in the electric field and consequently slow down. However, when moving towards lower field strength, the potential energy will decrease and the molecule will speed up again. Therefore, in a static setup no net deceleration is observed. The main idea of the Stark acceleration/decelerator is to switch the applied voltages from the electrodes so far used to create the electric field to the grounded electrodes and *vice versa*. When the electric field is switched while the molecule is within the field the molecule will lose the amount of kinetic energy that was changed into potential energy in the electric field. Consequently the molecule will slow down.

The simplest way to quantify the effect of switching the voltages is to define a synchronous molecule with velocity v_0 which will always have the same phase φ_0 with respect to the Stark energy when switching of the electrodes takes place. Hence, it will exactly travel the distance L in the switching time period Δt. Therefore, this molecule will always be at the same position with respect to the closest electrodes and will lose the same amount of kinetic energy per stage. Molecules with a slightly different phase φ or velocity will experience a correction towards to values of the synchronous molecule.[4] This ensures that molecules are kept together during the Stark deceleration [24]. The kinetic energy $\Delta K(\varphi_0) = -\Delta W(\varphi_0)$, that is lost per stage can be calculated at a given switch time by the difference in the potential energy at the positions $\pi z/L = \varphi_0$ and $\pi z/L = \varphi_0 + \pi$ and is given up to second order by [17]

$$\Delta W(\varphi_0) = W(\varphi_0 + \pi) - W(\varphi_0) = 2a_1 \sin \varphi_0. \quad (5.7)$$

[4] When for example a molecule has a larger phase than φ_0 it will lose more kinetic energy and will lack behind φ_0. Hence, in the next stage it will lose less kinetic energy. This will lead to an oscillation of the phase and velocity relative to the synchronous molecules.

Even though in the experiment the applied electrode voltages are switched between two static situations, an equation of motion requires continuous variables. Since the deceleration per stage is small compared to the kinetic energy of the molecule, we can assume that the deceleration of the synchronous molecule is due to a continuously acting average force $\langle F \rangle = -\Delta W(\varphi_0)/L = -2a_1 \sin\varphi_0/L$. In this picture the static switching between the fields is similar to a potential well that travels with $L/\Delta t$ and slows down the molecules (defined as traveling wave approximation). The motion of all non-synchronous molecules with the same velocity but a different phase $\varphi = \varphi_0 + \Delta\varphi$ as the synchronous molecule can be described relative to the molecules with φ_0. Subtracting the force corresponding to Eq. 5.7 for molecules with φ and φ_0 results in the equation

$$\frac{mL}{\pi}\frac{d^2\Delta\varphi}{dt^2} + \frac{2a_1}{L}\left[\sin(\varphi_0 + \Delta\varphi) - \sin(\varphi_0)\right] = 0, \tag{5.8}$$

which describes the motion of the molecules in a traveling potential well to a very good approximation [17]. When inspecting Eq. 5.7 it becomes clear that in the traveling wave approximation the operation conditions to decelerate molecules are $0 < \varphi_0 < 90°$, while for $-90° < \varphi_0 < 0$ the molecules will be accelerated. That this theory of the Stark decelerator is able to describe the experiments is demonstrated for the special case of $\varphi_0 = 0$ in Fig. 5.5a. For this phase, the Stark decelerator should only guide the molecules through the molecular beam apparatus. For OH radicals with an initial velocity of 450 m/s (Kr) the TOF profile with an intense peak at about 2.9 ms shown in Fig. 5.5a is measured.[5] This peak can be explained by taking the length through the molecular beam apparatus of 1.31 m (Fig. 5.4) into account and highlights that the molecules are only guided but not decelerated. The Monte-Carlo simulations employing Eq. 5.8 for the different low-field seeking states are shown in the same figure as gray lines (details for the simulations can be found in [17, 25, 26]) and nicely reproduce the experimental observations, when both low-field seeking states are taken into account. When now the phase of the experiment is chosen to allow deceleration of the radicals ($\varphi_0 = 50°$), the TOF profile shown in the upper part of Fig. 5.5b is obtained [27]. A peak at about 3.8 ms in the TOF profile indicates that the deceleration of the OH molecules was successful. For this particular case the molecules were slowed down from 465 m/s to 305 m/s. The fraction of molecules that are slowed down and are missing in the distribution of the fast molecules, is indicated by an arrow in all figures of Fig. 5.5b. Increasing the phase to $\varphi_0 = 70°$ slows down the OH radicals from 440 m/s to 170 m/s (Fig. 5.5b, middle part) or even further from 428 m/s to 21 m/s (Fig. 5.5b, bottom) when using $\varphi_0 = 77°$. Additionally an influence of the switched electric fields on all other not slowed down molecules becomes visible by the highly structured TOF profile (expanded bottom part of Fig. 5.5b). For more details see [17]. In the last mentioned case the radicals become that slow that they could be trapped with the help of electrostatic fields. For this

---

[5] The small sidebands in the TOF profile are molecules that are not trapped by the traveling potential well but experience the switched electric fields.

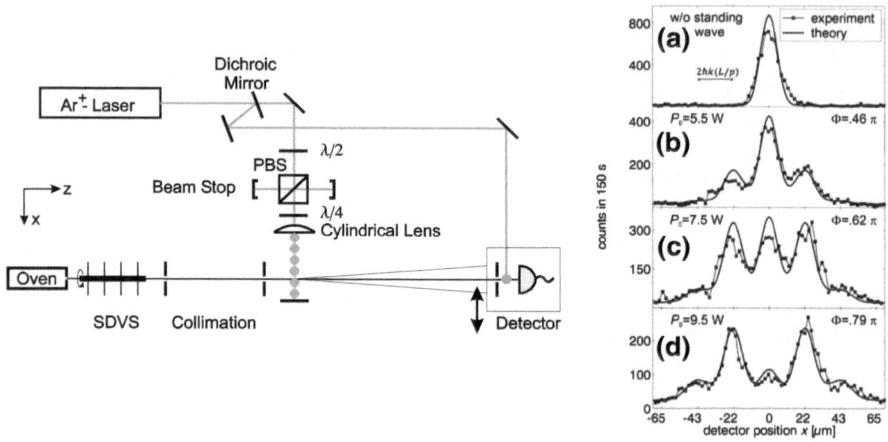

Fig. 5.5 **a** The none state resolved TOF profiles of OH with an initial velocity of 450 m/s (coexpanded with Kr) and $\varphi_0 = 0$ [23]. The radicals are only guided through the Stark decelerator resulting after 1.31 m in the intense arrival time peak at 2.9 ms. Consequently, the velocity did not change. Reprinted figure with permission from van de Meerakker et al. [23]. Copyright 2005 by the American Physical Society. **b** For $\varphi_0 = 50°$ (*top*), $\varphi_0 = 70°$ (*middle*) and $\varphi_0 = 77°$ (*bottom*), however, some of the molecules are slowed down to 305, 170 and 21 m/s, respectively [27]. The fraction of molecules missing in the distribution of the non-decelerated molecules is indicated by an arrow. For the case shown at the bottom an additional electrostatic trap was used, trapping the slowed down neutral molecules by using static electric fields of about −15 kV [27]. This results in an effective trapping of the molecules, as seen in the inset of the figure. Reprinted figure with permission from van de Meerakker et al. [27]. Copyright 2005 by the American Physical Society

purpose a molecular trap can be placed behind the Stark decelerator, allowing to trap the radicals while at the same time measuring their LIF spectra (for more details see [22, 27]). The effect of introducing such a trap on the TOF distribution can be seen in the bottom part of Fig. 5.5b. At about 7 ms a LIF signal is observed that reaches a constant value after some initial oscillations. This indicates the effective trapping of the OH molecules. Over a longer time period (Fig. 5.5b, inset) this intensity decays due to background gas induced losses of OH molecules. These results, therefore, clearly demonstrate that neutral molecules can be slowed down and trapped using their dielectric properties.

Similar experiments have not only conducted with OH [28] but also for NH_3, [22, 24] metastable CO [25, 29] and YbF [30] and many more small molecules [31]. A possible experimental application of the decelerated molecules is to perform scattering experiments. A crossed molecular beam experiment was performed investigating the inelastic collision of Xe with OH using a modified version of the apparatus shown in Fig. 5.4a [32]. The slowed down OH radicals are intersected by a beam of Xe atoms and the change of the state selectively recorded TOF profiles with and without Xe scattering was used to infer the inelastic scattering probability as a function of the OH radical velocity. Due to the variable velocity and the narrow velocity distribution very accurate inelastic state resolved low energy collision cross sections were obtained [32]. A comparison with theoretical modeling of the scattering process, furthermore, gave an excellent agreement with the experimental data, highlighting the possibilities that the Stark deceleration method offers. These experiments have just recently be extended in order to study state selective reactive scattering with very high resolution [33].

So far we have only discussed Stark deceleration (for which also other experimental realizations exist [34]) but other methods have been developed in order to manipulate the motion of neutral molecules. Very similar to the electric beam deflection studies are, for example, conformer selectors, that separate different conformers in space based on their differing dielectric properties enabling conformer selective studies [35]. Additionally other traps for neutral molecules have been developed which allow to trap the neutral species for several seconds [36]. In terms of focusing elements, it was demonstrated recently that a microwave lens allows to focus molecules in high- and low-seeking quantum states [37]. A much more detailed overview of all these experiments can be found in several review articles [17, 31, 35, 38]. However, all these results show that the manipulation of polar neutral molecules in the gas phase is possible, as long as their dielectric properties are well known. Therefore, the determination of the dielectric properties of isolated molecules and clusters is crucial to broaden the applicability of the herein described methods.

References

1. Nairz O, Brezger B, Arndt M, Zeilinger A (2001) Phys Rev Lett 87:160401
2. Knight WD, Clemenger K, de Heer WA, Saunders WA, Chou MY, Cohen ML (1984) Phys Rev Lett 52:2141
3. Tikhonov G, Kasperovich V, Wong K, Kresin VV (2001) Phys Rev A 64:063202
4. Davisson CJ, Germer LH (1928) Proc Natl Acad Sci USA 14:317
5. Schöllkopf W, Toennies JP (1994) Science 266:1345
6. Kadar-Kallen MA, Bonin KD (1992) Phys Rev Lett 68:2015
7. Kadar-Kallen MA, Bonin KD (1994) Phys Rev Lett 72:828
8. Ballard A, Bonin K, Louderback J (2000) J Chem Phys 113:5732
9. Gerlich S, Hackermuller L, Hornberger K, Stibor A, Ulbricht H, Gring M, Goldfarb F, Savas T, Muri M, Mayor M, Arndt M (2007) Nature Phys 3:711
10. Eibenberger S, Gerlich S, Arndt M, Tüxen J, Mayor M (2011) New J Phys 13:043033
11. Ulbricht H, Berninger M, Deachapunya S, Stefanov A, Arndt M (2008) Nanotechnology 19:045502
12. Hornberger K, Gerlich S, Ulbricht H, Hackermüller L, Nimmrichter S, Goldt IV, Boltalina O, Arndt M (2009) New J Phys 11:043032
13. Brezger B, Arndt M, Zeilinger A (2003) J Opt B: Quantum Semiclass Opt 5:S82
14. Hornberger K, Gerlich S, Haslinger P, Nimmrichter S, Arndt M (2012) Rev Mod Phys 84:157
15. Gerlich S, Gring M, Ulbricht H, Hornberger K, Tüxen J, Mayor M, Arndt M (2008) Angew Chem Int Ed 47:6195
16. Tüxen J, Gerlich S, Eibenberger S, Arndt M, Mayor M (2010) Chem Commun 46:4145
17. van de Meerakker SY, Vanhaecke N, Meijer G (2006) Annu Rev Phys Chem 57:159
18. Kroto HW (2003) Molecular rotation spectra. Dover Publications Inc., Mineola
19. Townes CH, Schawlow AL (1975) Microwave spectroscopy. Dover Publications Inc., Mineola
20. Brown JM, Carrington A (2003) Rotational spectroscopy of diatomic molecules. Cambridge University Press, Cambridge

21. Jongma RT, Rasing T, Meijer G (1995) J Chem Phys 102:1925
22. Bethlem HL, Berden G, Crompvoets FMH, Jongma RT, van Roij AJA, Meijer G (2000) Nature 406:491
23. van de Meerakker SYT, Vanhaecke N, Bethlem HL, Meijer G (2005) Phys Rev A 71:053409
24. Bethlem HL, Crompvoets FMH, Jongma RT, van de Meerakker SYT, Meijer G (2002) Phys Rev A 65:053416
25. Bethlem HL, Berden G, van Roij AJA, Crompvoets FMH, Meijer G (2000) Phys Rev Lett 84:5744
26. Scharfenberg L, Haak H, Meijer G, van de Meerakker SYT (2009) Phys Rev A 79:023410
27. van de Meerakker SYT, Smeets PHM, Vanhaecke N, Jongma RT, Meijer G (2005) Phys Rev Lett 94:023004
28. Wohlfart K, Filsinger F, Grätz F, Küpper J, Meijer G (2008) Phys Rev A 78:033421
29. Bethlem HL, van Roij AJA, Jongma RT, Meijer G (2002) Phys Rev Lett 88:133003
30. Tarbutt MR, Bethlem HL, Hudson JJ, Ryabov VL, Ryzhov VA, Sauer BE, Meijer G, Hinds EA (2004) Phys Rev Lett 92:173002
31. Schnell M, Meijer G (2009) Angew Chem Int Ed 48:6010
32. Gilijamse JJ, Hoekstra S, van de Meerakker SYT, Groenenboom GC, Meijer G (2006) Science 313:1617
33. Kirste M, Wang X, Schewe HC, Meijer G, Liu K, van der Avoird A, Janssen LMC, Gubbels KB, Groenenboom GC, van de Meerakker SYT (2012) Science 338:1060
34. Meek SA, Bethlem HL, Conrad H, Meijer G (2008) Phys Rev Lett 100:153003
35. Filsinger F, Meijer G, Stapelfeldt H, Chapman HN, Küpper J (2011) Phys Chem Chem Phys 13:2076
36. van Veldhoven J, Bethlem HL, Meijer G (2005) Phys Rev Lett 94:083001
37. Odashima H, Merz S, Enomoto K, Schnell M, Meijer G (2010) Phys Rev Lett 104:253001
38. Bethlem HL, Meijer G (2003) Int Rev Phys Chem 22:73

Chapter 6
Summary

The method to deflect molecules or clusters isolated in the gas phase by an applied electric field, has a longstanding tradition in physical and chemical science. Since the first discovery of Johannes Stark that energy levels can be influenced by electric fields (1913) [1, 2], over Erwin Wrede performing the first beam deflection experiments on alkali halides (1927) [3] to present day experiments on complex systems like metallic Na [4], molecular H_2O [5] or bimetallic clusters [6], electric beam deflection studies are a well established method for over the past 80 years. Despite the fact that dielectric properties have an outstanding significance for chemical and physical processes of neutral clusters and molecules, it is surprising that beam deflection studies (combined with near light force techniques for frequency dependent studies) are still one of the view generally applicable methods to deduce polarizabilities and permanent electric dipole moments from experiment.

Therefore, in the above presented chapters and sections we devoted ourselves to explaining the underlaying physical phenomena, experimental realization and theoretical background of the beam deflection method to every interested reader. Compared with the early beam deflection studies, the experimental accuracy, the general understanding of the experimental phenomena and the theoretical interpretation of the results of present day investigations have been improved successively over the years. Hence, position sensitive mass spectrometry or long deflection distances allow to record the mass and deflection information simultaneously or to investigate systems exhibiting very small mass weighted susceptibilities, respectively. Furthermore, the improvement and development of cluster sources [7–9] which give the desired particle densities and available ionization laser sources enable to study of complex cluster systems [10]. Additionally, two major factors have improved the interpretation of the experimental observations. First, due to working with cold cluster sources the experiments can be interpreted within the rigid rotor approximation allowing a simple and intuitive understanding of the experimental results using perturbation theory [11–13]. Only in the last couple of years it became clear that an even more in-depth interpretation is possible using classical or quantum mechanical simulations [14–16]. The second development in terms of interpreting the experimental

S. Heiles and R. Schäfer, *Dielectric Properties of Isolated Clusters*, SpringerBriefs in Electrical and Magnetic Properties of Atoms, Molecules, and Clusters,
DOI: 10.1007/978-94-007-7866-5_6, © The Author(s) 2014

results is the use of quantum chemical methods to find possible cluster structures and the corresponding properties. This then allowed, for the first time, to connect the experimental results with theoretical predictions for various cluster structures. Hence, a very careful analyses of the experimental observations to discriminate between different cluster isomers.

Beside all these improvements, still a large number of unsolved problems remain. For rigid clusters this especially influences asymmetric rotors for which under special circumstances a chaotic rotational motion was observed in experiment [5, 16–18]. However, more problematic is the influence of excited vibrational motions or isomerization processes on the interpretation of the experimental results. While experiments are readily performed for various temperatures of the cluster source, the effect of increasing the internal temperature of the cluster is yet not well understood.

In future beam deflection investigations, the successive improvement of the accuracy of cryogenic experiments and availability of interpretation models will allow to study not only the dielectric properties of isolated clusters in the gas phase but additionally learn more about their geometric structure. Especially, position sensitive mass spectrometer detectors [19] could enable to decrease experiment times and the accuracy compared to present day setups. Furthermore, an general understanding of beam deflection measurements at moderate temperatures could not only broaden the applicability of the method but additionally is a prerequisite to perform Stark deceleration experiments on more complex aggregates.

Acknowledgments The authors would like to thank all people that helped during the process of writing this manuscript. Very special thanks to Sascha Schäfer (now University of Göttingen) for establishing many of the herein described methods and many helpful discussions. Without the fruitful cooperations with Peter Schwerdtfeger and Roy L. Johnston most of the work presented would not have been possible. Therefore we wish to express our deepest gratitude to them and all of there group members. Furthermore, we would like to acknowledge financial support of the Deutsche Forschungsgemeinschaft (DFG SCHA 885/10-1). S. Heiles is deeply grateful for a PhD scholarship of the Fonds der chemischen Industrie.

References

1. Stark J (1913) Nature 92:401
2. Stark J (1914) Ann Phys 348:965
3. Wrede E (1927) Z Phys 44:261
4. Bowlan J, Liang A, de Heer WA (2011) Phys Rev Lett 106:043401
5. Moro R, Bulthuis J, Heinrich J, Kresin VV (2007) Phys Rev A 75:013415
6. Heiles S, Hofmann K, Johnston RL, Schäfer R (2012) Chem Plus Chem 77:532
7. Milani P, de Heer WA (1990) Rev Sci Instrum 61:1835
8. Moro R, Rabinovitch R, Kresin VV (2005) Rev Sci Instrum 76:056104
9. Lupulescu C, Rahim MAE, Antoine R, Barbaire M, Broyer M, Dagany X, Maurelli J, Rayane D, Dugourd P (2006) Rev Sci Instrum 77(1):125102
10. Matsuda Y, Bernstein ER (2005) J Phys Chem A 109:3803
11. Schäfer S, Assadollahzadeh B, Mehring M, Schwerdtfeger P, Schäfer R (2008) J Phys Chem A 112:12312

12. Bulthuis J, Becker JA, Moro R, Kresin VV (2008) J Chem Phys 129:024101
13. Bulthuis J, Kresin VV (2012) J Chem Phys 136:014301
14. Heiles S, Schäfer S, Schäfer R (2011) J Chem Phys 135:034303
15. Dugourd P, Antoine R, El Rahim MA, Rayane D, Broyer M, Calvo F (2006) Chem Phys Lett 423:13
16. Abd El Rahim M, Antoine R, Broyer M, Rayane D, Dugourd P (2005) J Phys Chem A 109:8507
17. Antoine R, El Rahim MA, Broyer M, Rayane D, Dugourd P (2006) J Phys Chem A 110:10006
18. Carrera I, Mobbili M, Moriena G, Marceca E (2008) Chem Phys Lett 467:14
19. Clark AT, Crooks JP, Sedgwick I, Turchetta R, Lee JWL, John JJ, Wilman ES, Hill L, Halford E, Slater CS, Winter B, Yuen WH, Gardiner SH, Lipciuc ML, Brouard M, Nomerotski A, Vallance C (2012) J Phys Chem A 116:10897

Index

S. Heiles and R. Schäfer, *Dielectric Properties of Isolated Clusters*, SpringerBriefs in Electrical and Magnetic Properties of Atoms, Molecules, and Clusters,
DOI: 10.1007/978-94-007-7866-5,